Plant Biosystematics

Plant Biosystematics

Prof. S. Venkat Ratnam

MD Publications Pvt Ltd
New Delhi
www.mdppl.com

Published by :

MD Publications Pvt Ltd
"MD House", 11, Darya Ganj,
New Delhi - 110 002
Phone : +91-11-41563325 (Customer Service)
+91-11-41562846 (Editorial)
E-mail (orders): order@mdppl.com
Website: www.mdppl.com

ISBN: 978-81-7533-160-0

Published and Printed by Mr. Pranav Gupta on behalf of **MD Publications Pvt Ltd** at Times Press, New Delhi.

PREFACE

Systematics can be used, as defined by Simpson (1961) as the scientific study of the kinds and diversity of organisms and if any and all relationships between them 1) Relationship of descent, 2) Relationships of similarity 3) Spatial or Geographic, and 4) Trophic relationship.

Most of the text books on the systematics emphasized the classificatory aspects of systematics and neglect to some extent the theoretical and evolutionary aspects of the field. Similarly most systematic books adequately cover the morphological and anatomical techniques used by the plant systematists but cover in much less detail the genetical, cytological, chemical and statistical approaches used in systematic work. The present text is to present the objectives of theoretical and technical aspects of evolution and proper understanding of evolution in detail. The book is designed to be used more of experimental aspects of the field of systematics, but also as a companion to any of the many fine plant taxonomy books being used in introductory courses.

Many systematists had been unhappy because existing classifications could not express all the relationships uncovered by systematics. They also felt that the accepted system of nomenclature did not allow a classification that expressed all that was known about the evolution of species.

Biosystematics is that part of systematics concerned with the **Variation** and **Evolution** of a species. It is often more concerned with the process of evolution than it is with the classification itself. Based mainly on the breeding systems several special classifications have been proposed.

Biosystematists have proposed a series of experimental classifications based on the genetic adaptation of the plant to its environment, particularly soil and climate, and its ability to interbreed with other plants both in nature and under experimental conditions. Turesson (1922) first formulated the ecotype concept, his classification was later modified and put on a more precise genetic footing by Clausen, Keck & Hiesey and by Gregor etc. Transplant experiments and breeding experiments have been the tools used in the construction of these special, experimental classifications. Failure to interbreed was recognized as of great evolutionary importance, since such populations are prevented from any further gene exchange now at any time when they meet in the future. So, biosystematists have therefore devised various ways of describing continuous variation by graphical or mathematical means.

Biosystematics as defined by Camp and Gilly was consequently an attempt to produce a system of nomenclature by which more relationships could be explicitly expressed.

The rate of evolution appears to have been frequently speeded up by aridity. Stebbins (1952) points out that there are numerous structures which plants can evolve in adaptation to dry conditions. These include the annual habit, bulbs, corms etc., reduction in leaf surface, succulence and indumentum.

There are about 50,000 kinds of plants living on the face of the earth. It is therefore imperative that our energies be directed toward the discovery and formulation of basic principles and mechanisms that direct the functioning and evolution of organisms, rather than toward a general survey. Such a selective approach is already prevalent in the more analytical fields such as molecular biology and cell biology, but is less prevalent in fields such as physiology, anatomy or systematics. In this book the attitude has been taken that the purpose of science is to establish general principles expressible in rigorous terms. These principles have to be related, however, to the outside world in a direct, precise, and meaningful way and the reader should satisfy himself that this is so before accepting any generalization. A number of specific examples are presented in the text to illustrate principles. No systematic review of the plant kingdom or any part thereof has been attempted. The most examples referred in this book are Angiosperms. Whenever the choice existed examples have been taken from the recent literatures. The purpose of this action is twofold, first to show the reader that systematics is a very active field and second to allow him to reach the original sources with the least effort.

Present day species are evolutionary products that have descended from preexisting species. Variation in organisms provides the raw genetic material that in concert with environmental variation, makes speciation possible. Variation in natural populations is limited by such factors as self fertilization, asexual reproduction, genetic fixation and stabilizing selection. Organic evolution involves the development of phyletic lines and the transmutation of form and structure within a line that gives rise to morphological species. The biological species represent the major continuity in ongoing evolution. Examples, therefore are not intended to give credit for the discovery of the principles they illustrate. As a matter of fact, no effort of any sort has been made to present the historical development of systematics except in the brief introduction.

Throughout the text book, the evolution theme " Genetic adaptation, or Organic evolution axiom is being continuously given an explanation in terms of organs such as phyllode, spine, sepal, petal, syngenesious anthers, petaloid anthers, their different orientation for promotion of cross pollination has been stressed.

Speciation is the development of freely interbreeding organisms adapted to their environment that are reproductively isolated from other such populations.

Finally the treatment of speciation includes the integral components of population breeding and asexual reproductive systems, isolating mechanisms and stabilization and selection along with a discussion of many types of species resulting from speciation. Plant populations are the source of potentially heritable genetic variation, which in concert with environmentally induced variation, leads to speciation.

Reproductive isolation comprises blocks to gene exchange between populations that stem from genotypically controlled differences in their reproductive organs, reproductive habits or fertility relationships.

The necessity or tendency for outbreeding may be reinforced by dioecism, self-compatibility, heterostyly, and other mechanisms. Out breeding is usually aided by the wind, insects or other pollinating agents. Out breeding generally maximize heterozygosity and variability are dioecism and monoecism, protandry and protogyny, heterostyly, self sterility and chasmogamy.

Introgression is a process of successive hybridization that cause the migration of genetic material from one species into another. Initial interspecific hybridization is usually followed by a series of back crosses to one of the parents.

I am greatly indebted to Prof V H Heywood (D Sc) and Prof D M Moore (PhD) and Late Prof J B Harborne (FRS) for receiving training under their able guidance.

I am immensely grateful to beloved Vice-chancellor, Prof Beela Satyanarayana for his consistent encouragement and unstinted support.

I express my sincere thanks to Prof V S Raju who has been source of inspiration.

I am very grateful to Prof R S Rao, who inculcated the Biosystematic ideas.

I would like to record and express sincere thanks to Mr J Lenin Bapuji for his diligent and expert typing of the manuscript.

Credits for figures have been scrupulously assigned in the accompanying legends although I have not been able to contact the authors for permission. To them whom my thanks are also due.

My final word of thanks is due to Mr Pranav Gupta, Director, MD Publications, New Delhi for his patience and kindness in author's sickness for seeing the book through press.

Prof. S. Venkat Ratnam

CONTENTS

Introduction and Historical Account

A question about nature that has no immediate answer in every plant science. When man began to wonder about the diversity of living organism's systematics started. The first step in a rational attempt to understand the diversity of an organism is to describe and classify them. The second step is to provide a plausible hypothesis for the systematized diversity and the last to devise ways to prove the hypothesis. The above mentioned three steps constitute the basis for the scientific method and mark the historical development of systematics.

The book deals with the principles and methodology of the part of biology primarily concerned with the diversity of plants and their interrelationship. The subject is variously known as taxonomy, systematics or biosystematics because each of these terms has special meanings, we shall first define them as they will be used in this book.

Taxonomy is the study of classification, its principles, procedures and rules (Simpson, 1961). Classification is ordering of plants into classes and groups. It is very basic, logical operation, because without some sort of initial ordering further comparative studies are made difficult if not possible. There are different ways of classifying and when the objects being classified are living organisms.

Nomenclature is the application of distinct names to each of the groups recognized in a classification. Certain very strict rules has been developed throughout the years. In a code of botanical nomenclature the rules are embodied that governs the application of names to plants. Classification, taxonomy, and nomenclature all deal with various aspects of the ordering of plants into some sort of logical system that tries to describe the diversity of living organisms.

Systematics can be defined as that of branch of biology that is concerned with the comparative study of organisms and all relations among them. There are four main types of relationships between plants 1) Relationships of descent, also called phylogenetic relationships, indicate the degree to which two individual plants are related to a common ancestor, 2) Relation of similarity, known also as phenetic relationships indicate the degree to which two individual plants look alike, not only in their external morphology, but, in all their other aspects such as anatomy, physiology and cytology, 3) Spatial or geographical relationships indicate how closely two individual plants are situated to each other and finally, 4) the so called trophic relationships indicate interdependence, that is, the extent to which two individual plants depend on each other, that is when one is parasite on the other or when both utilize the same mineral in the soil. Although systematists deal with all four of these relationships, phylogenetic and phenetic relationships emphasized. In order to ascertain the extent to which two plants are related, one must compare them. Comparison is, therefore, the basic approach of the systematics.

They can construct classifications in many ways but should be based on some kind of relationships among the organism being classified. Modern classifications of organisms are based on comparative studies, that is on systematic studies that include all relationships but emphasized phylogenetic and phenetic relations. The term biosystematy was first introduced by Camp and Gilly in 1943. They defined biosystematy as an attempt to delimit the natural biotic units (meaning the units of evolution) and to apply to these units a system of nomenclature adequate to the task of conveying precise information regarding their defined limits, relationships, variability and dynamic structure. What did they mean by this definition?

Many systematists had been unhappy because existing classifications could not express all the relationships uncovered by systematist. They also felt that accepted system of nomenclature did not allow a classification expressed all that was known about the evolution of species. Biosystematics as defined by Camp and Gilly was consequently an attempt to produce a system of nomenclature by which more relationships (phylogenetic relationship could be explicitly expressed).

Biosystematics is the field of study dealing with variation and evolution primarily experimental and analytical, and mostly treating the species and intraspecific taxa. The synthetic theory of evolution is the foundation upon which biosystematic rests.

Biosystematics is usually considered as the study of genetical and cytological phenomenon. Comparison, description and classification are the first operations that have to be performed in a biosystematic study. Explanation through the formulation of hypothesis is the next and providing the hypothesis experimentally the last. Actually description and explanation are integral parts of any scientific view of nature. The explanations given by the biosystematists are evolutionary explanations. That is, the relationships between plants are explained within the theory of evolution by natural selection.

Biosystematics defined as the application of genetics and cytogenetics statistics and chemistry to the solution of systematic questions in order to provide explanations about the diversity of organisms within the frame of the theory of evolution.

More vividly biosystematics have been analyzing breeding systems, studying patterns of variation, determining the evolutionary potential and doing the pertinent work in the chemical, numerical, cytological, anatomical, embryological, and palynological aspects of systematics for selected groups of organisms. This is the experimental or biosystematic state of development.

In this book major facets of plant biosystematics as here defined will be presented. In the first seven chapters some of the principal evolutionary problems that concern biosystematists and the questions that these problems pose will be analyzed. The various hypothesis that have been proposed to answer these questions will be presented as well as the evidence on which these hypothesis rest. In the last five chapters as the major experimental approaches of biosystematics will be discussed.

Historical Account

In Greece the roots of biology of western civilization are to be found. In addition to the knowledge about plants, rudimentary system of classification (Plants were classified into herbs, shrubs and trees), there was little transmitted to modern times in terms of correct specific botanical knowledge of wood cuts of tuberous plants were reversed so that tubers appeared as fruits or drawings of man eating plants were shown. The authors of these works were largely ignorant about the plants they presented in their books. The major concern of the herbalists was with medicine (Lawerence, 1965, Arber, 1928).Their approach was largely based on the superstition and preconception that the plant parts resembled the human organs that they could help cure.

These medieval forms of thinking were drastically changed in the sixteenth, seventeenth and eighteenth century as a reversal took place. This was so called Age of Reason and the analytical rational approach of the times, one that no longer looked at nature with preconceived ideas but with an open mind, was also felt in botany. Leonhurt Fuchs (1501-1566) who was professor of medicine in Tubingen, grew most of the plants that he illustrated in his herbals, observing their structures and properties rather than copying from earlier herbals as had been the prevailing custom. This resulted more accurate observations and a rekindling of curiosity about plants.

Of these two hundred years the rational approach of the Age of Reason produced great amount of information. First the native European plants and later plants from all over the world were grown in botanical gardens where they were studied and dissected. Chairs of natural sciences were established in several European Universities and basic aspects of plants were explored. With the work of the great Swedish naturalist Carolus Linnaeus (1707 - 1778) and the French school of botanist all these works were culminated who laid foundations of eighteenth and nineteenth century systematic botany.

Former Classifications Proposers

Linnaeus was the best known botanist and naturalist of his time. He described hundreds of new species of plants and animals sent to him by correspondents and disciples all over the world. He was a professor of practical medicine at Uppsala Sweden, and founded a famous botanical gardens still exist there today. Two of Linnaeus books, the *Genera plantarum* (1737) and *Species Plantarum* (1753) represent a compendium of the knowledge of the day.

In addition with the works in naming and classifying plants, Linnaeus reputation is based on two major accomplishments. The major one is the development of the system of nomenclature still in use today. Prior to Linneaus time it had been customary in the scientific literature to describe and name plants by short Latin sentences. The system was not most efficient one. Linnaeus developed a system of naming plants so that a plant carried the name of the genus to which it belonged followed by its own name. The name of the species was a one word adjective that described some characteristic of the plant. So, for example, the sunflower plant was named *Helianthus annuus* that is the annual species of Helianthus. This system was adopted because of its simplicity. But the adjectives chosen to describe the species were not always precise enough particularly as new species of the genus were described. For example today there are seventeen known annual species in the genus *Helianthus*.

Some later botanist, therefore changed some of the names given to the plants by Linnaeus and others so that they would be more precise. In addition in many instances plants were re-described and renamed by botanists who were ignorant of the fact that this has already been done. Indian corn was named *Zea mays* by Linnaeus, but in the next fifty years it was given more than fifteen other names, such as *Zea americana* and *Mayzea cerealis*. This naming and renaming created confusion, because clarity requires that an object has only one name. To alleviate this confusion practices, botanist defined toward the end of the nineteenth century to establish by common agreement certain rules on how to name plants. These rules state among other things, that the first name given to a plant is the valid one even if it does not accurately describe the plant, as in the case of *Helianthus annuus*. The rules make Linnaeus binomial system of naming of plants mandatory they further state that Linnaeus *Genera Plantarum* and *Species Plantarum* are the starting points of Plant nomenclature, that is names published prior to this time are to be ignored. The other accomplishment of Linnaeus is his sexual system of classification i.e, the properties of the flower circumscribe the major groups of plants. According to Linnaeus classification based on similarities of floral parts reflected more accurately than the phylogenetic relationship. The floral characteristics were considered more significant for classificatory purposes than vegetative ones until so proved. But later authors recognized the importance of all parts characteristic features that are useful.

Later John Ray (1627-1705) a graduate from Trinity College suggested a system of classification based on embryo characters and number of cotyledons. This was in consonance with the basic idea of delimiting plants on the basis of plant forms like trees or herbs. He recognized two major sub-divisions of flowering plants herbae and arbores. The former was subdivided into imperfectae (flowerless individuals) and perfectae (flowering plant individuals).

The later was divided into monocotyledons and dicotyledons. His meritorious system received appreciation with special reference to the recognition of genera *Baccifera* on the basis of pericarp character of fruit. But such preliminary and dynamic features like fleshy nature of pericarp were not very much reliable. He also made suggestions to subdivide Compositae into four classes. Undoubtedly he was trend setter for a natural system of classification of flowering plants John Ray's *Historia Plantarum* (1686-1704) described flowering plants under two main sub groups dicotyledons and monocotyledons.

John Ray's contemporary Pierre Magnol (1638-1735) subscribed a classification up to family level on morphological characters of root and stem. Floral and seed morphology was also used for the first time Genus *Magnolia* is commemorated after his name. One of Pierre Magnol pupil Joseph Pittonde Tournefort (1656-1708) professor of Botany at Jordin du Roi under Louis XIV made large collections of plants and published elements de botanique in 1694. This valuable work includes 10146 species spread over 698 genera. The illustrations of *Flora of Environ of Paris* (1698) and *Institutions Rei Herbarie* (1700) were his more important and distinction works, plants described in them are on copper plates. He also believed in broad sub divisions of flowering plants like herbs, shrubs and trees. One of the interesting merits of his system is description of several species under characteristic representative genera. He refused to accept existence of sexuality in plants. In his work, usage of relevant terms Petaliferous and Apetalous forms of corolla with separate and united petals or regular or irregular corolla is observed. He traveled widely in Europe and scaled **Arafat** Mountains. He died in an accident while crossing the street near the garden. Rudolph Camerasius (1667-1721) Director of Botanical garden at Tubingen, Germany, despite of being a non-taxonomist said that plants reveal sexuality and seeds were hither to produced only when pollen developed in stamens come in the vicinity of ovary and style representing the female sex. Without pollen, ovaries in flowers normally remained sterile.

His work is not evidenced by publication but he communicated his results to the reputed scientist and scholars at various universities. These findings provided a new conceptual outlook to approach the problem of classification.

In crowded galaxy of angiosperm classifications Bentham and Hooker system, Hutchinson's system, Takhtajan's system and Cronquist's system shine with a luster of their own and need a mention here as they represent climaxes of taxonomy. Admittedly Hutchinson promulgated a phyletic concept in letter and spirit. His classification appears closer to ultimate than others in following harmony with currently accepted information. Takhtajan's system is quite similar to Cronquist's for obvious reasons that both these excellent men developed their concepts independently but finally were in consultation with each other and the similarities of their views reflect necessities of current state of knowledge and considered nearest to the principles of morphological evolution. Cronquist's system is directed towards professional botanist or a reader of considerable insight in modern broad scale of taxonomy and clarity of philosophic concepts in the subconscious mind. Cronquist's classification despite being most recent and broad based awaits its approval and adoption in the institution of learning the world over. While the mind for fulfillment appetite tends to adopt Hutchinson's system of classification practically of the things cling to Bentham and Hooker's system the only one who describes Indian plant species adopt Hutchinson's system of classification.

Taxonomic study became inherent even before dawn of systematized knowledge. As plants inhabiting earth were most attractive and wedded to human civilization for their economic usefulness. It has known within the framework taxonomic principles that a natural system of classification should be perceived as best as possible and it was realized through the monumental work Origin of Species by Natural selection (Darwin 1859) that natural system reflects evolutionary relationships.

Many systematist had been unhappy because existing classifications could not express all the relationships uncovered by systematists. They also felt that the accepted nomenclature did not allow a classification that expressed all that was known about the evolution of species. Biosystematics consequently to produce a system of nomenclature by which more relationships could be explicitly expressed. Basically it represented the application of genetical, cytological, anatomical, palynological, phytochemical studies in particular in the description and circumscription of the basic unit of classification of the species.

G.H.M Lawrence used the term Biosystematics instead of Biosystematy which can convey precise information regarding their defined limits, relationships variability and dynamic structure, in other words, their classification as different orders of magnitude. Cytological studies provided critical data and were a dominant force in plant taxonomy for three decades. Biosystematics became associated with botanical cytogenetic studies. Along with cytogenetics, the data from anatomy, ecology, statistics etc were included in the taxonomic problems, particularly to the diversity of organisms within the frame work of the theory of evolution.

According to Davis Heywood (1963) biosystematics is a modern approach to taxonomy and an extended state of knowledge of the taxa. It is held as one of the stages through which taxonomy passes. Staflenu (1969) described it as evolutionary systematics of living organisms in the widest possible sense. Stebbins (1970) held that in contrast to classical taxonomy it concerns primarily with the process of evolution and secondarily with the end products of the same. It is genetical and cytogenetical approach to solve taxonomic riddles. Ross (1974) opined it as an investigational field of systematics, based on any scientific information that can be brought to bear on the problem of evolution, whether they concern speciation or phylogeny.

Raven (1976,1986) stated that the term biosystematics has outlived its usefulness and it is abandoned in favour of plant population biology and other equivalents. Munnoe (1977) concluded that biosystematics is an approach to systematics based on experimental, behavioural and genetical characteristics. While others used it for the whole array of indirect, theoretical, experimental and modern methods. Stace (1980) held that biosystematics is to convey virtually any taxonomic activity not pursued in the Herbarium or almost any newly aquired technique.

In taxonomy an individual plant is studied in (classical) taxonomy as a basic unit, in biosystematics a group of plants (population, species, and sub species) that interbreed each other having a definite geographical distribution in other words a population are treated as a basic unit. Thus the biosystematics is concerned with the study of genetic, evolutionary native of populations, or micro evolutionary phenomenon.

The data from cytogenetics, statistics, anatomy, palynology, embryology, phytochemistry are used to assess the species and infraspecific levels.

The term population genetics is concerned with the gene frequencies in population. It is a combination of genetics and systematics. Camp (1961) equated it with biosystematics, while Longlet (1971) to genecology. In practice, it is applied like genecology to interspecific studies. In which morphological differentiation is usually limited for formal classification.

Investigation of variation (diversity) through experiments in morphology, anatomy, embryology, cytology, genetics, palynology, phytochemistry, numerical taxonomy, molecular biology and the organic adaptation in terms of evolution, proposing the hypothesis on the evolutionary aspects of genetic adaptation to solve the taxonomic problems in terms of species, subspecies infraspecific and interspecific problems through breeding methods.

1

History, Scope and Importance of Biosystematics

Objective

1) To understand the evolutionary mechanisms at species level.
2) To study and understand the natural variation which is useful for adaptations.
3) To assess the historical development stages of systematics and biosystematics.

Introduction

The increase of knowledge about plants revealed interesting facts about vegetative world marked with relative progress of civilization. For the sources of food, shelter, clothing and medicine initiated hunt in old age. The ancient concept about plants has undergone an entire change with advance of knowledge and updated versions of published works. Plants provide an orderly and comprehensive array of curiosities for their identification and categorization. For obvious reasons these problems became crucial before botanist. The morphologically, biochemical and genetical aspects with ramifying ancillaries in many other domains of science spread with the modern knowledge of science. Taxonomy is one of the branch which comprehends all others. The cumulative historical advance has gradually terminated into various systems of classification of plants. There has always been an effort to place forms showing resemblances nearer in a system of classification in view to provide a natural position to plants. Phylogenetic considerations also gained importance. The history of taxonomy is fascinating and full of controversies and is still reshaping itself systematized story of plant description dates back to the period of Bock, de Obel, Brunfels and Fuchs, Philosophical giants like Theophrastus, Aristotle, Dioscorides and Pliny added more scientific and systematic approach to the subject. Dioscorides published valuable work the "material and *Anicia Juliana*". These comprises complete description of plants with coloured illustrations. Plants are our fundamental sources of food and energy, of shelter and clothing, of drugs and beverages, of the oxygen we breathe and form the aesthetic base of our environment.

The ecologist studying pollutant decomposers has to have the organisms which are involved identifying the genetist introducing new germplasm into a crop, for increased production or disease resistance has to know the characteristics of the source of plant and the chemist analyzing plant material for cancer inhibiting drugs has to know the name of organisms from which the material was procured. Related decomposers may prove to be more efficient removers of pollution; closely related species may be a source of better germplasm for increased production, genes more resistant to diseases and other genotypes may provide cancer inhibiting drug that will be effective with fewer side effects. The scientific name is communication medium to which all information is attached. He has to device simple system of classification to show relationships and breeding mechanism with which genes are not polluted. Hence these biosystematic activities are basic to the operations of man.

Biosystematics is a dynamic science. The duties of bisystematist are never ending. The constant stream of information, revisions of identification schemes, reevaluation of classification systems and perception of new relationships or increased effectiveness in communication about plants and long as plant world exists, there will always be more to learn about plants, plant products and plant taxa for both practical and theoretical purposes.

The great challenge today is how to preserve the plant life of the world. We need to learn more about the organisms, which are basic to our survival, and how to store and retrieve the vast amount of information on plants for man's use is primary duty, and one of the major problems.

Such a useful subject which comprises all aspects of plants from various subjects requires the tracing of the historical account.

Historical Developmental Stages

Pliny (23-79 A.D) was a Roman naturalist who described medicinal plants exhaustively. His work is spread all over 37 books. The progress in taxonomy was gradual and assisted by all those interested in various aspects of commercial plants.

Mainly the earlier descriptions were for medicinal purposes. Particularly cultivated plants were studied from such view points in China, Assyria, Egypt and Pre Columbian Aztee culture of China. The taxonomic considerations were confined to a narrow range of family, genera and species and largely assisted by plant breeding work. Later on concept to delimit wider groups of herbs, shrubs, and trees led to consider close similarities between individuals as an important feature in taxonomy rather than differences. Accordingly suggestions were made for creation of wider groups rather than generic and specific ones as Rushes, umbellifers, and grasses.

16th Century Advancement

Andrea Caesalpino (1519-1603) an Italian physician

Director of Botanical garden at Bologna used Latin and Italian words for nomenclature of plants. His collection of 768 plants were preserved as relevant herbarium in Natural history museum at Florence. He published 16 voluminous books under the title **'De plants'**. They include description of 1500 plant species without illustrations and synonymy.

John Ray (1627-1705) from Trinity college, Cambridge suggested a system of classification based on embryo characters and number of cotyledons. With this clue the basic idea of delimitation of plants are on the basis of plant forms like trees or herbs. According to this forum two major subdivisions of flowering plants are classified: herbae and arbores. The former was subdivided into flowerless and flowering plants (perfectae). The flowering plants were divided into monocotyledons and dicotyledons. John Ray was trend setter for a natural system of classification of flowering plants in his '**Historica Plantarum**'.

17th Century Advancement

The great Swedish botanist and naturalist described hundreds of new species of plants and animals sent to him by several disciples from all over the world. Two of Linnaeus books the '**Genera Plantarum**' (1737) and '**Species Plantarum**' (1753) represent a compendium of knowledge. Linnaeus developed a binomial system of naming plants which became popular and acceptable. The names published prior to this time are to be ignored. Linnaeus also recognized that the floral parts were more useful than the vegetative organs in grouping like with like.

John Baptiste Pierre Antoinede Monet (Claevalein de Lamarck (1744-1829)) was the earliest of the French naturalist exposing this view with any clarity. According to Lamarck the species changes in time and space based on the habitat.

During the last part of eighteenth and beginning of nineteenth centuries formal relationships between species and natural affinities were not established.

Until the beginning of the nineteenth century, scientific botany was largely systematic botany even though pharmacology was the most popular aspect of botany at that time. After Linnaeus time the automization of the field into a series of subdisciplines began. Morphology, anatomy, cytology, plant physiology, embryology, all developed as independent fields during the nineteenth century. The perfecting of optical instruments, the increase in total knowledge and the increased professionalization of science that accompanied the industrial revolution contributed to this development.

Charles Darwin (1859) profounded most intellectual theory of "Origin of Species" by Natural selection contributed to the idea of evolution, but ignored the inheritance mechanism.

Gregor Mendel (1822-1884) was the first to propose the principles of inheritance in 1865, but his work was ignored until inheritance mechanism was rediscovered at the beginning of the twentieth century. The fusion of Darwinism ideas with genetic principles has given rise to the modern theory of evolution.

Origin of Synthetic Theory of Evolution and Biosystematics

With the addition of the knowledge and concepts of Paleontology and systematics it has developed into the so called synthetic theory of Evolution. The synthetic theory is the foundation on which biosystematics rests, which has altered the direction of taxonomic research, because the diversity of organisms can be not only described but also explained in many instances with a firm basis of theory. In Darwins time very little was known about the facts of reproduction and inheritance. These developments came during the twentieth century and had a profound effect on the methodology of systematics and led to biosystematics as shown in figure 1.1.

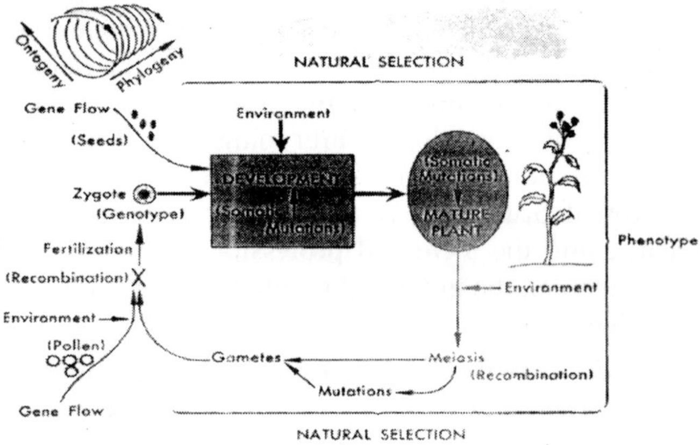

Fig. 1.1. Schematic representation of how the major forces of evolution interact. A zygote, the first cell of the organism, is the result of the fusion of two gametes. It will develop into a mature plat. Through the complex and little-known process of development (Indicates in the sketch as a "black box"). The various environmental factors will determine which of the genetic potentialities of the genotype are finally expressed in the phenotype, the mature plant. Natural selection is manifested in that some genotypes (zygotes) will die during development because of their genetic composition, environmental factors, or both, whereas other genotypes will produce sterile phenotypes. Or phenotypes with reduced fertility. The mature plant proceces in turn the gametes that will fuse with other gametes to produce a new generation of zygotes. During meiosis and fertilization new combinations of genes are produced, the process known as *recombination,* that result in novel genotypes. Genes can mutate at any stage in the cycle; however, only those mutations that are transmitted through the gametes will have any lasting effect on evolution. Finally through the introduction of pollen and seeds from other populations. Gene flow takes place. Every complete cycle is known as *ontogeny;* a series of ontogenies is a *phylogeny.*

Impact of Darwinism

Darwinism had great impact on the species concept. It produced two major concepts of biosystematic interest are the concept of descent with modification and the concept of the breeding population. Both of these concepts can be tested experimentally at the species level. At higher levels in the taxonomic hierarchy these concepts can be applied only by inheritance.

Descent of Species with Modification

The concept of descent with modification implies that heriditary process is crucial in determining similarities and differences. It is not surprising then that genetic experiments play a major role in modern systematics. The concepts of the population means not that idealized type represents the species, but that all individuals do consequently, statistical concepts also play a major role in biosystematics. Finally, modern biology has demonstrated conclusively the chemical nature of life. Chemistry is beginning to play an increasingly important role.

Summary of Historical Account of Plant Systematics and Biosystematics

Gathering more accurate information about plants for erecting a system of classification is primary objective. This progression culminated in eighteenth century with Linnaeus, who provided a system of nomenclature and a system of classification that for a short while was accepted almost universally as "Natural System" of the later part of the eighteenth and the nineteenth centuries as well. Here the emphasis was on systematic studies that took into account all aspects of the plant. The relationships which was uncovered by the systematists earlier ushered in a new phase was provided by Darwin's theory.

In this century new experimental approaches particularly genetics, cytology, chemistry, and statistics permit systematic studies encompass many more aspects of the biology of the plant than before, and these approaches also allow a certain degree of experimental work. Molecular genetics, nuclear chemistry have been utilized in offering solutions.

Scope of Biosystematics

Now a days modern equipments such as NMR (Nuclear magnetic resonance), Spectrophotometer. Atomic absorption spectrum, Automatic amino acid analyzers, protein analysis. Automatic analysis of secondary metabolites, DNA sequencing for comparative study of different species of a genus and family, Nucleotides, Nucleic acids, Semantides, Isozymes, Allozymes, RFLP, RAPD analysis have been and will be utilized in establishing relationship and evolutionary significance. With ultramodern computer the morphological, anatomical, chemical, cytological data can be processed within minutes for deducing the significance to offer solutions to solve the taxonomic riddles. The bioinformatics studies will broaden the biosystematic research.

Definition

Camp and Gilly (1943) coined biosystematy and defined it as variation and evolution.

1. To limit the natural biotic evolutionary units
2. To apply to these units a system of nomenclature adequate to the task of conveying precise information regarding their defined limits, relationships, variability and dynamic structure.

Theme of Biosystematics

Many systematists had been unhappy because existing classification could not express all the relationship of nomenclature which did not allow a classification that expressed all that was known about the evolution of species. However Biosystematics has represented the application of genetical and cytological criteria in systematics studies and in particular in the description and circumscription of the basic unit of classification of the species with statistics.

Clausen *et al* (1945) defined biosystematics who regard genetics, cytology, comparative morphology and ecology as furnishing the critical data which together when applied to the study of organic evolution makes up biosystematics.

Biosystematics can be defined precisely as the application of genetics (cytogenetics), statistics, and chemistry to the solution of systematic questions in order to provide explanations about the diversity of organisms within the frame of the theory of evolution.

The Kinds of Relationships has to be Established

1. Descent relationship or phylogenetic relationship
2. Similarity relationship or phenetic relationship
3. Spatial or geographical relationship
4. Trophic relationship or parasite host.

Role of Biosystematics

1. Biosystematics provides information for all branches of biology and broadening of all biological sciences.
2. It is indispensable for economic medicinal plants and gives vivid picture of the existing diversity on earth.
3. All aspects of biology will be worked out in biosystematics and full information will be made available to the other branches of sciences.
4. Mathematics is a mother all physical sciences - Biosystematics is mother of all biological sciences.
5. Generate informational system.
6. Biosystematist acts as conservators.

Central theme of Biosystematics

Biosystematics is the field of study dealing with variation and evolution; biosystematist analyze the breeding systems, studying pattern of variation, determining the evolutionary potential and doing pertinent work in the chemical, numerical, cytological, anatomical, embryological, palynological and molecular aspects of systematics for selected group of organisms. This is the experimental or biosystematic stage of development.

2

Biosystematics in the Framework of Evolution

Objectives

i) To assess and understand various definitions of biosystematics.
ii) To understand the natural and other selections.
iii) To understand the genetic sources of variation.

Various Definitions of Biosystematics in the Framework of Evolution

Camp and Gilly (1943) coined a term biosystematy and defined it as variation and evolution.

Evolution is also defined as variation and natural selection, and adaptation.

Biosystematics in its broadest sense, is that part of systematics concerned with variation and evolution of species. It is often more concerned with the process of evolution than it is with classification itself.

Understanding of Natural Selection Process

The principle of natural selection is cornerstone of theory of evolution and must be well understood if modern biosystematics is to be comprehended. Natural selection, the fundamental law of evolution is the basic mechanism that makes the evolution possible. It was enunciated by Charles Darwin and Alfred Wallace at the same session of the Linnean Society of London July 1st 1858.

Natural selection

The preservation of favourable individual differences and variations and the destruction of those variations which are injurious, is called *natural selection* or survival of fittest - Darwin. Natural selection is the evolutionary force that tends to produce systematic and heritable change between one generation and the next which may result in adaptation and survival, variation and speciation or extinction. Directional selection occurs when the environment is changing in a systematic fashion, leading to a regular directional change of adaptive characteristics of breeding populations.

In its most elementary formulation, natural selection is based on the fact that gives two or more units (genotypes population species) that are capable of self reproduction, the one that produces the largest number of offspring surviving to reproductive age will be more abundant in the following generation.

One popular definition of natural selection (Lerner, 1958) is the non random differential production of genotypes. The term fitness (Darwin fitness) is used to designate the ability to produce fertile offsprings that survive to reproductive age. The best fit unit (genotype, population) is the one that produces the largest number of reproductive offsprings.

The ways by which a large number of reproducing offsprings produced are varied and knowledge of them is important to an understanding of the evolution of plants. Basically there are only two major ways for an organism to increase fitness. 1) by increasing the total number of offspring i.e., by increasing the reproductive capacity 2) by reproducing offsprings, that are better adapted to the environment and therefore have a greater probability of survival. Under circumstances, one or the other of these two mechanisms, becomes but basically natural selection tends to maximize both reproductive capacity and adaptative natural selection. "The preservation of favorable individual differences and variations and the destruction of those which are injurious, I have called natural selection or survival of the fittest." - Darwin.

Definitions of Evolution in Genetic Terms

In genetic terms the natural selection is the differential reproduction of genotypes- Lerner. Most specifically organic or biological evolution is responsible for the variation and change in the organisms through time.

Evolution

A directional change in gene frequency, over time, in response to selective factors of the environment; the continuous genetic adaptation of organisms to their existing environment.

The differential reproduction and survival of genotypes without the conscious intervention of man called as natural selection.

Source of genetic variation

From the stand point of view of genetics, evolution is the change in the frequency of the genes of a population. In order for a change to occur there must be at some time a source of new genes. The sources are of two types - 1) Mutation 2) Introduction of genes from neighbouring populations in the form of seeds or pollen.

Furthermore, in order to have change, there must be a force producing change, the force is of course, natural selection, although occasionally change occurs solely as a result of chance, a phenomenon known as genetic drift.

Genes do not occur by themselves, they are integral part of the chromosomes of the plant. Likewise, natural selection does not act directly on genes, but it affects phenotypes. Through the process of the sexual reproduction (recombination), the genes of a population are shuffled and reshuffled in each generation into novel combinations of genotypes that give rise to new kinds of phenotypes. The process of recombination, although it does not give rise to new genes, plays a very important role in evolution through the production of new phenotypes. As a matter fact, it is so important that recombination not mutation gives rise to most of the variation that is observed in population.

Recombination is a reassortment of chromosomes, cross over segments, and therefore genes that produce new genotypes in sexually reproducing organisms. Recombination is the immediate source of variation observed in populations (such as height of plant and colour of corolla) and is the raw material on which the process of evolution and speciation act. Factors controlling recombinations in plants (Grant, 1981) are operative at meiosis (chromosome number, frequency of crossing over, hybrid sterility) and fertilization (breeding system, population system, dispersal system, population size, crossability barriers, and external isolating mechanisms).

Types of Selection

There are three main forms that natural selection can take place -

1. Stabilizing selection
2. Disruptive selection and
3. Directional or progressive selection.

These three types of selection represent ways by which the population remains adjusted to an environment, which may either -

1) *constant and stable (stabilizing selection)*
2) *becoming more variable and breaking up into different sub environments (disruptive selection) or*
3) *changing constantly in a single direction (directional selection)*

The consequence of selection is usually though of as phenotypic change. This is not necessarily so and as a matter of fact, often it is not so. Actually, a very common type of selection, is stabilizing selection. In stabilizing selection, phenotypes with characteristics near the mean of the population are more fit than those at the extreme on account of stable environment. As a result of this kind of selection, the mean of the population will not vary although its variability may, depending on the genetic determination of the character under consideration.

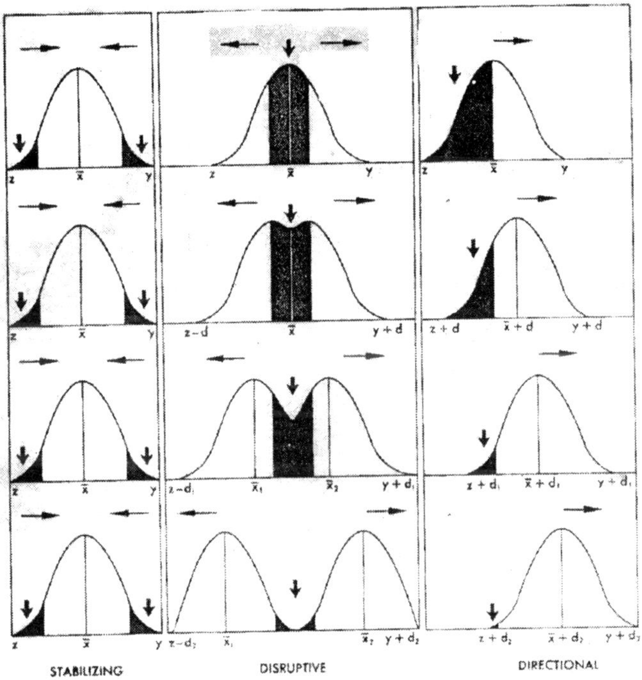

Fig. 2.1. The three main kinds of selection. In stabilizing selection the environment favors the organisms with values close to the population mean, consequently little or no change is produced in the population; disruptive selection favors the extremes, and will tend to divide the population in two; directional selection favors one extreme, and it will tend to move the mean of the population towards that extreme. Directional selection accounts for most of the change observed during evolution. In the figures the curves represent the frequency of plants with a certain range of values between x and y; the shaded areas, those phenotypes that are being eliminated by selection; the long arrows, the direction of evolutionary change; d=amount of change.

Stabilizing Selection

Blooming period in many spring blooming plants is under stabilizing selection. Members of spring blooming species that bloom too early in the season will, on average produce fewer seeds than those emerging later that introduces genetic variability at a slow but steady rate whereas natural selection is the conservative and leveling force that tends to standardize all the individuals of a population at a peak of maximum fitness. In as much as most mutations decrease fitness, these two forces work against each other as we have just seen.

Attaining Genetic Novelty and Variability

The sexual reproduction creates new diversity, when the gametes that combine to form the zygote are genetically different. This diversity can be assumed to happen when the gametes come from different organisms. When the combining gametes come from the same organisms, it may not produce diversity since gametes of homologous organisms are all genetically alike. For this reason sexual reproduction can act efficiently to increase or maintain variability only in genetically heterogenous cross-fertilizing populations.

The mode of reproduction also determines to a large extent the type of variability of species based on genetic novelty. The genetic recombination and gene mutations and mutations will bring forth the genetic novelty evidently leads to morphological variability.

Biosystematists have therefore derived various ways describing continuous variation by graphical or mathematical means. Huxley (1938) proposed the term cline for a character gradient i.e., any variational trend in space appearing in a population or series of populations of a species. A cline on a geographical or regional basis has been distinguished as a topocline and a local one related to ecological gradient i.e., an ecocline.

Gene Flow and Intensity of Selection

Gene flow is the introduction of new genes into a population from outside sources, usually continuous populations, in the form of occasional crosses between individuals of different populations as a result of the introduction of foreign seed or pollen. In these cases, genes derived by mutation and maintained by selection in a different environment are introduced into the population. The immediate effect is similar to the effect of mutation a new source of variability in the form of new genes that are introduced.

The effect of gene flow is that of cohesive force within the species that can work counter to selection to certain extent, depending on the rate of gene flow and intensity of the selection, and that can prevent neighbouring population from becoming two different population. Because of the effect of gene flow which is directly related to the number of immigrants and intensity of selection, and the farther apart two populations are physically and the more their environments differ and the less this cohesive forces it felt.

The probability of genes staying or being eliminated in a population is determined by its frequency, which is the result of mutation, selection and gene flow. Random fluctuations that are the result of selection that can occur in populations from year to year. These will have seldom long lasting evolutionary effects, but they will keep the population from reaching its theoretical highest fitness. Random phenomenon can be of evolutionary significance only when a number of important gene arrangements are completely eliminated from the population because pollinating insects have not yet emerged, plants blooming late in the season will on the average produce fewer seeds than those that bloomed before, because of lack of water or because of competition with other plants.

Another kind of possible selection is one that favors the extremes over the centre on account of an environment that is breaking up into two or more types. But because the extremes breed with each other in each generation, a large number of organisms having intermediate characteristics will be produced.

Disruptive Selection

In an environment favoring the extremes any mechanisms that would keep opposite extremes from breeding with each other would be highly advantageous, leading to second type of selection, called *disruptive selection*. Disruptive selection requires some kind of mechanism that ensures like breeds with like or in other words, some kind of isolating mechanism.

Directional Selection

A final kind of selection is *directional selection*. This is the type of selection where one extreme phenotypic expression of character is favored; which happens in nature when the environment is changing constantly in one direction. Directional selection produces phenotypic and genotypic changes more rapidly than any other kind of selection.

Directional selection probably accounts for the majority of the phenotypic changes that occur during evolution. However any kind of selection produces genotypic changes by direction. If no phenotypic and / or genotypic changes occur, or if such changes as occur are truly random, no natural selection exists.

Gene Flow

Mutation and selection are phenomenal that occur within the population and are not dependent on events occurring elsewhere. Mutation is a random phenomenon. It is the ability of different organisms to exchange genetic information with each other, the process the geneticist acclaim. The selective advantage of sex is so direct and so great that we can assume it and it arise at a very early stage in the history of life. Genetic recombination can speed up evolutionary change enormously and assist in emancipation from the environment.

Numerous mechanisms evolved in due time to make recombination increasingly precise in every respect. How the variation is generated in plants and animals.

In diploidy two homologous chromosome sets, one derived from the father, the other from the mother, of an elaborate process of meiosis during which homologous chromosomes exchange pieces, so that the chromosomes of father and mother are transmitted to the grand children not intact, but as newly reconstituted chromosomes with a novel assortment of genes. These mechanisms regulate genetic recombination among individuals by far the major source of genetic variability in higher organisms.

In addition to genetic recombination followed by independent assortment certain gene mutations and chromosomal mutations will bring variation.

Gene mutations include, DNA nucleotide mispairing the base analogous elements will be exchanged or base elements will be transmutated and this brings about genetic mutations.

The chromosomal mutations include, deletion or deficiencies, additions, inversions and translocations that will bring about some change in the architecture of the chromosome and evidently bring about the variation.

In some plants and animals one or two chromosomes will be lost or gained. In diploidy plants if the above mechanism occur resulting nullisomics and trisomics.

Polyploidy Drop

In a situation where the environment is very severe the chromosomes in the hybrid will be doubled. In the doubled chromosomes, certain chromosomes behave erratically and a pair with three other chromosomes forming associations of four such situation leads to the sterility. In that particular situation the plant may loose or gain one or two chromosomes resulting aneuploidly, polyploidy or otherwise called *polyploidy drop*.

Chromosomal Polymorphism

In this way variation will be generated. In certain vegetative propagated plants mostly the chromosome number exhibit variation in number in the same tissue of a species or different species of the same genus generating chromosomal polymorphism and such phenomenon also generate variation. Variations are raw materials upon which natural selection or selection pressure will be operating to generate useful adaptations to adapt the organisms or adapt themselves to the existing environment. Evolution through the genes is the way through which stable adaptations will come about. In this way the different individuals of a species adapt to different environment organ by organ to survive in their existing environment. In one environment phyllode, another spine in still another tendril in this way organic adaptations are brought into the organisms via genes by selection pressure which operates on the variation generated through various mechanisms.

Experimental Classification Based on Genetic Adaptation

Biosystematists have proposed a series of experimental classification based on the genetic adaptation of the plant to its environment, particularly soil and climate, and its ability to interbreed with other plants both in nature and under experimental conditions.

Progressive Adaptation is Real Evolution

The evolution is progressive adaptation, and therefore that all the morphological differences between organisms discerned by taxonomists and used by them in the construction of classifications are no more than by-products of the process of becoming better adapted. The study of morphological variations took on a new significance in relation to adaptation and survival. The twentieth century saw the introduction of new experiments and observational methods of studying variation, methods which began to throw light on the actual mechanisms of evolutionary change.

3

Relationships of Orthodox Taxonomy and Biosystematics

Objectives

i) To understand and assess the aims, system of classification and taxonomic structure of Biosystematics, Classical taxonomy.

ii) To have comprehensive understanding of the source of data, tests of characters and methods of description.

iii) To get clear cut idea of concepts of natural selection.

iv) To understand the similarities between alpha and omega taxonomy.

Note: capital letter indicates classical taxonomy, small letter indicates biosystematic character.

Aims

A) To describe all existing kinds of plants, to classify them according to their resemblances and differences and to name them according to a body of internationally agreed rules.

a) To identify evolutionary units and by experiment to determine their genetical inter relationships and the role of the environment in their formation.

Unit of study

B) Individuals or died dried specimens, twigs or small number of individuals.

> **b)** In asexual organisms, the biotype, in sexual organisms the breeding population or large representative sample of it. Material preferably living population.

System of Classification

C) Basic unit- The 'species' which is accepted as having objective reality (originally because of the dogma of special creation and later through the interpretation of evolution as simply another word for the origin of species)

> **c)** Don't give much importance to the classification where classifications are produced they are intended mainly to present in a systematic fashion data which have been obtained about natural populations. The basic unit adopted are diverse, according to immediate purpose and the system in use, e.g., ecotypes, coenotypes, conviva.

Taxonomic Structure

D) A hierarchy of categories embracing the whole of the plant kindom and named according to the internationally agreed rules.

> **d)** No general system is adopted, but in some groups minor hierarchies have been constructed to reflect the genetical relationships between populations. None of these extends upwards beyond the level at which genetical experimentation becomes impossible.

Source of Data

E) Inherent in the plants a) determinable from dead material external morphology, later internal morphology b) from living material little taken into account except features like periodically flowering time, deciduousness etc. Non-inherent geographical distribution.

e) Source of data inherent from plants, from dead materials, from living materials, observation of complete life cycles. Study of physiological behaviour and morphological adaptation in different habitats. Genetical tests of interfertility, cytological studies aimed at establishing breeding systems, etc. Non inherent geographical distribution, ecology.

Test of Characters

F) Intuition (based on accumulated experience trial and error method)

f) Observational techniques using statistical methods, experimental techniques using transplanting to test phenotypic variability and controlled crossings to determine the modes of inheritance.

Methods of Descriptions

G) Based upon individuals and the type concept. The giving names accepted as an important process.

g) Based upon populations or population samples using statistical methods. Nomenclature matters considered of little importance.

Concepts of Natural Variation Involved

H) Essentially static- assumes continuity of forms within species except for small amount of "sporting" and variation between individuals. The adaptative significance of differences between plants is not taken into account. The taxonomic hierarchy is accepted as suitable form of representation of natural variation, and its application in the internationally agreed form is regarded as obligatory in all groups, family, genus, species.

h) Essentially dynamic- full weight is given to the importance of the genetical systems. (i.e., breeding behaviour) operative in the groups under study. No rigid species concept is accepted or required. The internal variability of sexual populations is recognized, and the adaptative nature of much population differentiation is acknowledged and made a deliberate object of the study. Classification is based upon evolutionary units where these can be recognized and is not committed to a set of formal categories. No general classification of plant kingdom is attempted.

I) Orthodox taxonomy can ignore the new ideas about natural variation from genetical, cytological, physiological, ecological and other research subjects for the last century.

i) In biosystematics they will get the new ideas about natural variation from genetical, cytological, physiological, phytochemical, embryological, ecological variation data for the solution of taxonomic questions pertaining subspecies, variety and races.

J) Higher units than species are based exclusively upon degree of morphological resemblance or dissimilarity, although phylogenetic ideas have led to the construction of putative character series intended to show lines of descent.

j) The genus is represented in different environments that is occupied by groups of populations physiologically adapted to these environments.

These groups are arranged into species, which are separated by genetic barriers of various distinctions. Such species are ecospecies and few cases are of monotypic (coenospecies) usually containing one ecotype.

The genetic basis of the species is of more importance than the particular combination of morphological characters it brings forth.

When Mayr (1959) says that the subspecies category is not suitable for describing the population structure of the species (which is the aim of many biosystematists) he does not suggest that it should be abandoned as a classificatory unit. On the contrary, he notes that although the subspecies is not a unit of evolution but conceals much of the inter and intra population variability of species.

Comparison of Biosystematics and Classical Taxonomy

Biosystematics	*Classical Taxonomy*
1. Methodical empherical investigation of variation.	1. Basically a descriptive science.
2. Arrives at conclusion based on the experiments in cytology, genetics, anatomy, palynology and seed coat sculpture and phytochemistry.	2. Arrives at conclusion on the basis of grass morphological characters derived from the specimens.
3. It is experimental taxonomy.	3. It is descriptive science.
4. The knowledge from ecological and phytogeographical sciences is correlated with morphological characters selected are used to set boundaries or criteria.	4. Identification and names of plants only.
5. Aimed at systematic conclusions based on cytogenetics, cytotaxonomy, comparative anatomy, embryology, and other chapters which are correlated with morphology.	5. Taxonomist aimed at systematic conclusions based on pure morphology.
6. Omega taxonomist number has increased.	6. Taxonomist in alpha school has diminished in number.
7. Make use of available pertinent evidence to the solution of taxonomy.	7. Relies mostly on morphology.

8. Biosystematists get a training in cytology, genetics, anatomy, morphology and other branches of sciences also.	**8.** They have only training in description.
9. Seek collaboration with other fields of science.	**9.** Confining only to morphology.
10. Making a revision for all the families of all genera and prepare a correct monographs.	**10.** They prepare only floras based on the regions are vague not compared with world floras.
11. Classification will be extended to further two units such as subspecies, micro-species and varieties.	**11.** Classification ends up at the species level.
12. Devised various ways of describing continuous variation by graphical and mathematical means.	**12.** No graphical and mathematical describing.

Biosystematics is Basis or Mother of All Biological Sciences

Mathematics which forms the basis for physics, chemistry, meteorology and engineering subjects. So also the biosystematics and taxonomy forms basis for other branches of botany as mother subject. Some of the native Indian plants were patented by others due to lack of correct monograph on each family. But people are fringed to work on biotechnology without knowing correct epithet. Therefore, patentation was done by other country persons.

4

Biosystematics-Biosystematic Categories

Introduction

The different populations of the same species occupying distinctive habitats frequently showed morphological and physiological traits, often of an adaptive nature. These differences were due to modifications of the same genotype as a result of direct environmental influences or to the expression of different genotype combinations in different habitats. When grown in uniform conditions exhibit persistent differences.

General Objectives

1. To provide the student with an understanding of evolutionary mechanisms at species level.
2. To develop a working knowledge of field and laboratory methods of measuring, interpreting and presenting the data on natural variation, which is the **'raw material"** of organic evolution at the individual, population and intra population levels.
3. To acquaint the student with methods of experimental design in evolutionary problems.
4. It reveals numerous interesting evolutionary

phenomenon aiding to the other branches of biology.

5. To understand the natural relationship among the species of a genus.
6. To draw correlation based on the variation measured by other branches of biology
7. To strive to determine objectively whether a taxon belongs to the category of genus, species, subspecies etc.

Introduction to Development of Biosystematics

To cover newer approaches in Taxonomy, Camp and Gilly (1943) proposed some aims under the heading ***Biosystematy***. Lawrence (1965) was uneasy with the above word and recoined it as Biosystematics which is accepted form scientifically. Claussen, Hiesey and Keck (1945) Lawrence (1965) Valentine and Love (1958) Solbrig (1968) Grant (1971) and several others developed the subject. According to them classification and categorization of variation studies play an important role in which one need.

1. To delimit the natural biotic units (eg. species, genus)
2. To apply to these units a system of nomenclature adequate to the task of conveying precise information regarding their defined limits, relationship, variability and dynamic structure.

Kinds of Relationship

Many systematists had been unhappy because existing classification could not express all the relationship uncovered by systematist.

Biosystematics studies include all the relationships but emphasises on phylogenetic and phenetic relation. These relations include four types.

a. Relation of descent, also called phylogenetic relationship.
b. Relation of similarity known as phenetic relationship.
c. Spatial or geographical relationship.
d. Tropic relations including parasitism, commensalisms etc.

In this approach emphasis is placed on cytogenetics and cytotaxonomy supplemented by the classical approaches of morphology, ecology and phytogeography ; wherein cytogenetics is a combination of cytological and genetical techniques in the effort to arrive at the solution of taxonomic problem.

Methods of Study of Biosystematics

1. The first step in biosystematics is an investigation or a thorough sampling of the taxon and its populations and the cytological study of the chromosomes of many populations within geographic races, species genera and so on. Difference in chromosome number, their morphology and behaviour at meiosis usually indicates genetic differences or similarities of taxonomic significance.
2. The determination of the ability of the different populations to hybridize, and to study the different population which hybridizes, and to study the vigour and fertility of the hybrids which may discloses the presence or absence of natural breeding barriers between groups.
3. In meiotic behaviour of hybrids the homologous of chromosomes will be determined and pairing behaviour indicates the degree of genetic relationship in the material. The cytological data obtained from different geographical populations and races to establish fertility relations by making some crosses among geographical races and to study meiotic behaviour of homologous chromosomes of obtained hybrids will be compared with the data obtained from comparative morphology and geographical distribution.

When it is applied has an increased objectivity over one obtained through a consideration of morphology and distribution alone. According to Claussen (1945) the data obtained from genetics, cytology, comparative morphology and ecology when applied together to the study of organic evolution make up biosystematics.

Different Definitions

Solbrig's Definition

i) Biosystematic as an application of genetics (cytogenetics) statistics the chemistry to the solution of systematic questions in order to provide explanations about the diversity of the organism within the frame of the theory of evolution.

ii) Investigation of variation through experiments in comparative morphology, comparative anatomy, seed coat microsculptural features, reproductive biology, phytochemistry, embryology, karyotypic studies, meiotic behaviour studies and to correlate these characters with adaptation of organisms existing in that locality; and to describe the diversity of organisms to delimit it to assess the evolutionary adaption processes which occur within the population. Further to provide a plausible hypothesis for the systematized diversity for the sake of devising ways to prove that hypothesis is called as biosystematics.

Biosystematic is concerned with the variation and evolution of species. It is concerned with the process of evolution than it is with classification itself.

Mechanism of Study of Biosystematics

Biosystematic studies include the investigation of plant variation and to differentiate between the genotypic and phenotypic variation patterns and which way that variation will be used for better adaptation. It also indicates about the evolutionary limitations by apomixis and vegetative propagation and hybridization how polyploidy especially allopolyploidy, amphiploidy aneuploidy and polymorphism will promote the evolutionary mechanisms. Different geographical and ecological places certain physiological heritable adaptive characters will be produced in an organism which will be selected by either biotic or abiotic selection.

The Species can be recognized morphologically but the

characters merely serve as markers for the boundaries determined by the genetic methods. Mophological evidence is used to recognize species which are defined in genetic terms. Instead of creating morphological boundaries around species, geographical, ecological adaptational characters, changed chromosomal data of cytological and genetical evidences are used to create a sound boundaries around species.

Biosystematists are analyzing breeding system, studying patterns of variation, determining the evolutionary potential and doing the pertinent work in the chemical, numerical, cytological, anatomical, embryological and palynological aspects of systematics for selected groups of organisms. When Mayr (1959) says that subspecies category is not suitable for describing the population structure of the species which is the aim of many biosystematist. He does not suggest that it should be abandoned as a satisfactorily unit. Although subspecies is not a unit of evolution but conceals of the inter and intra population variability of species.

The experimental taxonomist strives to arrive at; an overall picture of biological relationships by attempting to correlate the relationship evidenced by ecology, cytogenetics, (gross micro-morphology,palynology) Phytogeography and Physiology. There are advantages and limitations to each of the approaches to a common goal, but it is significant that the findings of Taxonomy (classical as well as modern) and of biosystematic one of mutual importance and that each contributes to the other to build a stronger single disciplines. Lawerence further identified biosystematics with experimental taxonomy. Some equated the study of population genetics with experimental systematics. The experimental systematist usually begins with classical interpretation of species and works backwards so as to understand the genetic mechanisms of his material. The population geneticist begins with the raw population, discarding in his own mind any classical concept of species and works from there into a series of group concepts which may or may not be comparable to the taxonomist concept of species. Experimental Taxonomy does not mean the use of experimental procedures but is the taxonomic position of organisms in respects of its population rather than individuals, and the evolutionary process which occurs within populations.

Based on the data of cytology, genetics, ecology, and morphology the biosystematist developed a classification for experimentally investigated natural taxa and such categories are not intended as substitutes for the units used in classical or practical taxonomy. They are not equivalents, but they may be counterparts of them and which do not have nomenclatural status; each category provides a single term for a biosystematic situation. The category terms should be applied to either to a plant or population only after they have been experimentally proved to exist their stand. None of the category terms should be applied to a plant or population unless its right to the category has been established and recorded.

Objectives of Biosystematic Categories

1. How different individuals of the same species acquired the physiological, morphological adaptive characters ameliorated and fixed in their genotypes in different geographical regions.
2. Based on their crossability and gene exchange among the individuals how the respective category will be established.
3. How each category is equated with the classical taxonomic - counterparts in hierarchial category.

Introduction

Seeds and propagules of different plants are migrating to different geographic localities. For the sake of better adaptation each individual ameliorates itself by producing heritable physiological, morphological adaptive characters with which it will be surviving when they are brought to the experimental gardens, tested for their phenotypic expression and tested for their gene exchange through crosses. Thus each category will be established, they will be equated with the taxonomic counterparts in hierarchical category. First type of category indicates intraspecific categories among the individuals of same species or genus.

Turesson suggested the ecotypes would appear to be useful as a designation for subspecific segregates of which the salient characters are ecological rather than morphological.

Biosystematic species (Grant, 1981) are classified as categories based on fertility relationships as determined by hybridization experiments. These biosystematic categories form a hierarchy of ascending units; ecotypes, ecospecies, coenospecies, comparium and subspecies (Turesson, 1922, 1925, Clausen, Keck and Hiesey, 1939, 1945). It is often forgotten that this was a genecological, not simply a genetical hierarchy as is clear from the original definition (Turesson, 1922, Heywood, 1959). An **ecotype** consist of all the members of species that are fitted to survive in a particular kind of environment with in the total range of the species (Clausen, Keck and Hiesey, 1945). Two or more ecotypes are able to exchange genes freely without loss of fertility or vigour in offspring (Clausen, Keck, and Hiesey 1945). The Ecotype is the basic unit in biosystematics. It is accepted as a phyletic unit, adapted to a particular environment but capable of producing fully fertile hybrids with other ecotypes of the same species. Ecotypes are individual plants of a species adapted to different geographical habitats and exhibit heritable physiological adaptations even in the uniform experimental gardens and can exchange genes freely without loss or vigour or fertility.

Nature and Origin of Ecotypes

Ecotypes do not originate through sporadic (fortuitous) variation preserved by chance isolation. They are on the contrary to be considered as effect of the habitat factor, products arising through sorting and controlling. The ecotype concept implies that some population fractions can be demonstrated to be adaptively related in respect of certain features to a particular environmental factor within the species range. The basis of genecology, as Harberd (1958) says is the observation that this-intra-specific variability can be demonstrated to be of genetic origin and is not randomly dispersed throughout the species range, but is distributed in such a way that neighboring plants tend to resemble one another. The adaptation can be demonstrated in morphological or physiological characters.

In ecotype differentiation adaptation refers to any feature of an organisms or its parts which is of definite value in allowing that organism to exist under conditions of its habitat (Daubenmire,1959). It is recognized, however, that all the factors of the environment are interrelated and contribute to the selective forces operating on natural populations. A single genotype may be able to grow successfully under a wide range of conditions, often assuming different morphological forms. The adaptability of the race lies in its ability to grow under a wide range of habitat conditions.

Since not all genotypic differences between population will be adaptive. Environmental selection of genotypes most fitted to a particular habitat, and selective elimination of unfitted genotypes giving genetically adapted, heritable ecological product is the ecotype.

The ecotype is an experimental category, so that its adaptive nature to be proved by experiment. The different results of environmental selection have both been referred to loosely as *ecotypes*, it has been suggested that the genetically adapted fixed situation be distinguished as a genecotype. The situation where the flexibility that is genetically determined and parallels the genecotype in its response to environmental conditions might usefully be termed as *phenecotype*. Under conditions of geographical isolation as on islands, so called insular races may be built up and be considered as geo-ecotypes which have not been formed as the result of selection.

Populations which are morphologically, physiologically and developmentally adapted to live under different environment or habitat conditions are ecotypes.

Turesson put forward three basic propositions to explain the origin of ecotypes.

A. A species with wide ecological amplitude shows variation in morphological and physiological characters in different areas.

B. The differences are related to habitat variation.

C. The differences are not due to plastic response to change in environmental but are actually due to natural selection to locally adapted populations from a pool of genetic variations available in the species.

Difference in ecotypes, therefore, result from natural or discriminating selection of biotypes by unlike habitats. Therefore, the wider the ecological amplitude of a species the more numerous are its ecotypes.

Since the ecotypes are inter fertile they produce hybrids where the range of two ecotypes overlap. Outside this region the ecotype exist unaltered. The hybrids between ecotypes are, however, incompletely adapted to habitats of their parents, therefore their number is never large. By chromosomal doubling certain ecotypes are formed polyploidy often leads to the formation of new ecotypes. Since polyploidy seldom exhibits same ecological tolerance as their parents, loss or addition of chromosomes segments also produce changes in genotypes leading to the formation of new ecotypes(polyploidy, aneuploidy).

Significance

Formation of ecotypes helps the species to external ecological range and spread to new areas. Cultivation of economically important plants in climatically and edaphically different places has been made possible due to evolution of new ecotypes. It is done by planting the number of ecotypes and their hybrids in an area. One or the other ecotypes shows adaptation to the new habitat. Morphological variations can be marked in the species growing on varying habitats which lead to evolution.

Recognition and Classification of Ecotypes

Any feature of an organism or its parts which is of definite value in allowing that organism to exist under the conditions of its habitat, is called Ecotypic differentiation of adaptation.

The Ecotype status of population depends primarily on habitat recognition. A relation need to be established between the population heredity and the prevailing habitat factors. **Ecotypes** are therefore recognized as the product of the selective action of the whole environment, but they are defined with particular environmental characteristics, As Gregor puts it, one is forced to select that part of the heredity which is related to certain, more, conspicuous features of the habitat. The factors other than these may had a part in the pattern of variation observed need to be remembered.

Three main sets of Ecotypes are normally recognized, climatic, edaphic and biotic depending on which of the environmental factors has been the most dominant some other kinds of ecotypes have been described.

1. Climatic ecotype (Turesson, 1930) syn. Climatype,eg. *Leontodon, autumnalis*
2. Edaphic ecotype. (Ramakrishna, 1961, Gregor, 1942) eg. *Euphorbia thymifolia*
3. Biotic Ecotype. (Sinskaja, 1931). eg. *Portuluca oleracea* phototeriodic ecotype (Kaul, 1959). *Xanthium strumarium.*
4. Geographic ecotype syn. Reclussion type (Turesson, 1927) geo-ecotype (Gregor, 1931) eg. *Hieracium umbellatum.*
5. Climatic–edaphic ecotype eg. Cenchrus ciliaris (Panday and Jayanth, 1970).

One must bear in mind that selection operates not on single characters but on character complexes, even though one or few characters may be found to show variation related to the chosen environmental variable. Overemphasis on single characters may, as Stebbins points out (1950) confuses rather than clarify our understanding in the variation pattern. The ecotypic fraction of a species or population in response to one selected feature of the habitat may not coincide with its subdivision in response to another which became major source of difficulty in the taxonomic treatment of genecological differentiation. Taxonomic recognition will concentrate more on this individuality than on the demonstrable ecotypic picture, and in case of ecoclinaly distributed variation, taxonomic treatment is bound to be unsatisfactory, as it will weight the characters which contribute to the individuality of the local by differentiated population and in so doing cut across clinal patterns (Heywood, 1959). The difficulties of demonstrating variants have been considered by Bradshaw (1959) and by Wilkins(1960). It may be morphological expression of an underlying physiological difference.

Ecotype and Taxonomy

When adaptive variant or ecotype show correlated differences influenced by morphological treatment, they may be considered eligible for taxonomic consideration following normal taxonomic principles. The means by which morphologically different populations are detected: provided they are on a scale considered worth recognition. This can be attempted after ecotypes are found to coincide with recognized taxa for example in *Orchis latifolia, Geranium robertianum Silene dioica, Potentilla glandulosa* or in other cases taxa have to be errected to accommodate them. Claussen, keck and Hiesey originally considered the morphologically separable climatic ecotype as the basis of subspecies.

Many difficulties of which the principal ones will now be considered

1) The adaptive features of the ecotype do not lend themselves to taxonomic treatment or not correlated with features which do in several cases.
2) Most morphological differences between ecotypes are quantitative, depending on polygenic inheritance, they may require statistical treatment for their detection and it may be found that the variation within the ecotypes

obscures their racial differences.

3) Genecological variation in populations may frequently occur at lower level than formal nomenclature classification can usefully do. **Orthodox taxonomy** at micro evolutionary level has severe limitations and a problem that is seldom considered is the need for taxonomic recognition of ecotypic variation (Heywood, 1959).

4) Daubenmire (1958) states that recognition of ecotypes taxonomically is for ecological purposes. He further emphasizes that it is clearly useful for variants to have names, so that they can be easily referred to.

5) The presence of many species in floras is common to them will suggest a degree of similarity which may be highly misleading, if the plants belong to different ecotypes (Curtis 1959).

6) It became very difficult to recognize the micro taxonomic units below the subspecies level. The subdivision of the species all the way down to the **forma** levels was practiced in pflanzenreich monographs. The lower units are too small to occupy the serious attention of taxonomist. But many practicing taxonomists find need for the occasional use of all the infra specific units for dealing with the variation in well studied groups.

7) Cytological behavior karyotypes and their behavior are observed in different forms. The differences in cytological behavior show existence of distinct ecotypes.

Ecospecies

All the ecotypes so related that they are able to exchange genes freely without loss of fertility or vigour in the offspring. Turesson defined **ecospecies** as a group of plants comprised of one or more ecotypes within the coenospecies, whose members are able to interchange their genes without any harm to the offspring. Related ecospecies are usually separated by incomplete genetic barriers, which in addition to ecological barriers, are adequate to present free interchange of genes with any other ecospecies. When ecospecies of one coenospecies are crossed, the resultant hybrids are either partially sterile or if fertile, they produce many weaklings. In F_1 generation (mostly misfit) such weaklings are unable to compete, and fail to reproduce. During segregation of few such hybrids may possess adequate vigour to survive. Some related ecospecies likely inhabit different but often nearby ecological or geographical areas, thus maintaining relative genetic purity. The ecospecies almost nearly equal to the conventional and conservative taxonomic species.

Coenospecies

All the ecospecies so related that they may exchange genes among themselves to a limited extent through hybridization this is equivalent to the biological species.

Coenospecies of the same comparium are separated by genetic barriers that all the hybrids between them are sterile unless amphiploidy occurs. The coenospecies is a group of plants representing one or more species of Common evolutionary origin as far as morphological, cytological and experimental facts indicate coenospecies of the same comparium are separated by genetic barriers, that all hybrids between them are absolutely sterile, unless amphiploidy occurs. That is why distinct coenospecies may exists in a single environment without genetic intermixing

The Coenospecies are parallel to the taxonomic sections or subsections of the gonus. Here ecological separation does not play part in this definition. Like that of the ecotype, ecospecies, the identification of a phyletic unit as a coenospecies must be based on genetic experiment to determine the degree of fertility.

Comparium (Syngamodeme)

All the coenospecies between which hybridization is possible either directly or through intermediaries. Gene exchange between coenospecies is impossible.

All of those coenospecies which can be crossed through any of their members to produce viable hybrids corresponds to the comparium of Dansereau. The comparium marks the theoretical level beyond which the evolutionary history of different lineage must become finally independent.

The comparium is the biosystematic unit that often is comparable to the genus. It is composed of one or more coenospecies that are able to intercross and complete genetic incompatability prevails between them.

Among the different coenospecies that are incapable of gene exchange may produce hybrids and these will initially sterile may regain fertility by chromosome doubling may thus be added allopolyploids with the potentiality of further evolutionary development. According to Stebbins (1950) biosystematic categories and concepts involve also the use of the genetic terms of genotype, phenotype and biotype.

Biotype and Phenotype

Each individual organism consists of genotype and phenotype. The total outward (external) appearance is called phenotype. The actual total expressed characters or observable total form of an organism is called phenotype. **Genotype** combines with environment and makes what we can observe about the individual its external morphology, anatomy, behaviours, functions sum total of the attribute it reveals during its life have the phenotype. In some occasions individuals of the same genotype may appear to be different (i.e they have two different phenotypes). The phenotypic structure may alter in structure in response to change in the environment, these alterations are sometimes being partially reversible.

Genotype

Defined as the total genetic component of organism or in different sense to a specific allele, in a particular individual. Hereditary constitution of an individual given by its ancestors are represented by two sets of genes present in the zygote.

Biotype

Biotype consists of all the individuals having the same

genotype.

The biotype in cross fertilized organisms usually consists of a single individual. But in self fertilized plants, the individuals may become completely homozygous and produce by selfing a progeny of individuals all within the same genotype and so belong to biotype .

Turesson (1922) and Dansereau (1929) proposed two separate schemes, In Turesson system there are three ranks Coenospecies, ecospecies, ecotype. In Dansereau system the definition of the terms are such, not any one of the six is synonymous with the other (comparium, commiscum, convivium) Dansereau's lowest term convivium covers the definition of both ecospecies and ecotype; while his higher terms comparium and commiscum define the higher and lower aspects of coenospecies, Turesson employed a lower term ecophene, that denotes the ecological variant purely effected by the environmental modification of phenotype. Clements used the term **ecad** for such variation.

Subspecies

When the patterns of variation had become evident as consequence of the extensive collections of materials and it is intraspecific category which is geographical segments of a species. So subspecies would show a certain degree of morphological intergradation between themselves. A geographical subdivision of a species deemed worthy of formal recognition by a taxonomist.

A cline may be defined as a group and progressive change in the characters of a species when it is traced along an extensive tract of natural distribution. Since the variations are extensively the extremes often appear to be sufficiently distinct to be recognized species.

In localized studies, a species can easily be separated from its sympatric taxa by a clear gap. This was the kind of species recognized by Ray and Linnaeus and represent the species as it exists at a particular place at a given time. Such species are called non dimensional species (Mayr, 1969). The populations comprising such species might fall in discrete groups enabling the classification into subspecific group. Huxley (1939) classifying a series of integrading populations which are called clines.

These subspecies categories have been accepted by many on taxonomic grounds while some others have thought of them only in positional terms. In taxonomy, we proceed both upwards and downwards from species the basic unit, and hence the order species, subspecies, varieties are placed in the descending order. Evolutionary taxonomists Amadon and Short (1976) felt that it is desirable if not essential to have a category for the treatment of subspecies approaching species status. Short (1969) however suggested that the term be restricted to recently evolved allospecies.

Subspecies is defined as a monophyletic group of entirely or essentially allopatric species that are too distinct to be included in a single species. Amadon (1966) definition reads 'A group of entirely or essentially allopatric taxa that were once races of a single species, but which now have achieved species status.

Deme Terminology

Deme Terminology was used with a prefix, denoting any group of individual of a specified taxon. The nature of the association of the group of individuals or the kind of population unit which they may represent must always be specified by the addition of an appropriate prefix, second order, compound prefixes may also be used to specify relationships more precisely. The word, deme used alone simply means a group of individuals of specific taxon and shows no other relationships unless this is specified in the context (Davis and Heywood, 1963, Heslop-Harrison, 1967) Various basic deme terms are indicated below: the deme terminology equivalents are also indicated where appropriate in connection with other concepts described in this section.

A new system of deme terminology was proposed by Gilmour and Gregor (1939) and developed more fully by

Gilmour and Heslop-Harrison, (1954).

Deme is denoting any group of specified taxon (orthodox categories) can be used as frame work into which other classification for a particular purposes can be fitted to provide an infinite flexible series of categories which could be used to define any group of individuals on the basis of any set of criteria. This deme terminology system in its original concept, is non-hierarchial and it falls outside the scope of formal taxonomic categories (genus, species, etc). This concept avoids the use of root words such as species and its types which are associated with the latter.

The main purpose is the use of natural root, deme which implies nothing in itself except a group of related individuals of a particular taxon. The precise meanings of the terms are provided by various prefixes of which only three were originally proposed.

Topodeme: a deme occurring with in a specified geographical area.

Ecodeme: a deme occurring within a specified kind of habitat.

Gamodeme: a deme composed of individuals which interbreed in nature.

Later Gilmour and Heslop-Harrison (1954) expanded these suggestions and provided seven further basic terms.

Phenodeme: a deme differing from others phenotypically.

Plastodeme: a deme differing from others phenotypically but not genotypically. denoting reproductive behaviour

Genodeme: a deme differing from others genotypically.

Autodeme: a deme composed of predominantly self fertilizing (autogamous) individuals.

Gamodeme composed of individuals which are so situated spatially and temporally that within the limits of breeding system all can interbreed.

Endodeme: a deme composed of predominantly closely inbreeding (endogamous) but dioecious individuals.

Agamodeme: a deme composed of predominantly apomictic (non-sexually reproducing) individuals.

Clinodeme: a deme which together with other such demes forms gradual variational trend over a given area.

(***Cytodeme:*** a deme composed of individuals of all with the same karyotype).

Cytodeme: A population differing in some distinctive cytological features from others, Heywood, (1965).

Cytodeme: a deme which differs from others cytologically (Rad, 265).

Role of Biosystematics

1. It supplies almost exclusively the information needed for all branches of biology.
2. It is indispensable in the study of economically or medically important organism.
3. It contributes significantly to a broadening of biology and to built a better balance within biological sciences.
4. It gives out for us a vivid picture of the existing organic diversity of our earth and is only science that does so.
5. The empirical discovery that molecules can be evolutionary clocks, has been applied to a variety of problems in evolution and systematizes.
6. Mathematics forms the basis for all physical sciences whereas biosystematics forms the basis for all biological subjects and provides an accurate information.

SUMMARY

Comparison of the Biosystematic Categories of Turesson (1922), Dansereau (1929) and Gilmour (1939) and Heslop- Harrison (1954)

	Turesson	Dansereau	Gilmour	Heslop - Harrison
Individuals occupying a particular habitat and forming an interbreeding population which differs genotypically from other such populations produce fertile Hybrids.	*Ecotype*	*Convivum*	*Genecodeme*	Group capable of hybridizing with other groups to give hybrids showing complete fertility.
Individuals capable of exchanging genes freely without loss of fertility or vigour in the offspring the resultant hybrids are either partially sterile, if fertile they produce many weaklings, the weakling are unable to compete.	*Ecospecies*		*Hologamodeme*	Group capable of hybridizing with other groups to give hybrids showing some fertility.

Individuals capable of hybridizing with one another produces sterile hybrids unless amphiplody occurs.	*Coenospecies*	*Commiscum*	*Coenogamodeme*	Group capable of hybridizing with other groups but such hybrids are sterile.
Individual incapable of hybridzing with one another complete genetic incompatability prevails between them.		*Comparium*	*syngamodeme*	Group incapable of hybridizing with any other such groups.

1. Biosystematics may therefore be considered as the taxonomic application of the discipline known as genecology. The study of the genotypic and phenotypic variation of species in relation to the environments in which they occur

2. The subject broaden the application of taxonomic characters to all groups of plants.

3. The compilation of inventory of plants down to the species level and incorporating them in monographic rather than floristics.

4. The application of biosystematic data used to formulate plans of action in the conservation of genetic diversity.

5. New Ecotypes, ecospecies, Coenospecies of a species enable if to be adapted to climatically and edaphically to different places extended ranges of areas.

5

POPULATION SIZE AND DENSITY VARIATION IN INDIVIDUALS

Objectives

1. To understand the developmental, environmental induced and genetic variation
2. To assess discontinuous variation that results in discrete units that can be recognized, classified, circumscribed or named.
3. To illustrate the chance fluctuation in gene frequency in sexual reproduction that leads to heterozygous nature.
4. To examine how the heterozygous nature leads to speciation since population are units of speciation.

Introduction

Population is becoming a key word in taxonomy. In sexually reproducing organisms the evolutionary unit is the local breeding population. Population is unit of classification of Biosystematics. Populations which were sampled statistically give a valid representation of their variation. In population the averages indicate ***the frequencies of gene pool of gene alleles and chromosomal architectural variations***.

Definition

Population is a group of individuals of same species living and sharing the same habitat and climate for that period. Plant populations are the source of potentially **heritable genetic variation**, which in concert with environmentally induced variation leads to speciation. It is **genetic axiom** that the larger the interbreeding population, the greater the size of the gene pool. A large gene pool provides the genetic variation in a population necessary for its adaptation to changes in the selective forces of the environment. A small population contains fewer different alleles in its gene pool for adaptive response to changing environmental conditions. The number of individuals at a new site is controlled by the number of propagules per unit area as well as by the stability and spatial dimensions of the habitat for germination, establishment, and maintenance of the organisms. The fate of the population may well depend on the initial size and heterozygosity of its constituent individuals. A few populations started by a single seed in all probability will produce a very limited gene pool of genetically uniform individuals with low speciation potential. Morphological and physiological differences between populations are frequently associated with environmental differences, and appear to be adaptive. These differences have been studied extensively by genecologists and biosystematists but have been stunned by Taxonomist. Current needs in ecology and genecology make taxonomic treatment more necessary.

Population Size - Gene Pool

Population size includes more than one gene pool. It involves some form of competition between the individuals of the group. Generally, the greater the number of individuals per unit area(density) the greater the competition between the organisms for moisture, nutrients and light and more severe the natural selection. Density directly affects pollination and the reproductive capacity of the population.

Population *versus* Suitable Habitats Available

The number of populations within an area is determined by the number of suitable habitats available for colonization. Some species are restricted to a very specific soil types, lithology or set of moisture conditions. The nature and availability of dispersal agents determine the pioneering characteristics of certain species. For the black willow (*Salix nigra*) to produce new populations low water sand bars have to be present for germination and establishment almost immediately after dispersal of the seeds from the mature tree. If the sand bars are scattered and small, the black willow populations will be scattered and small, assuming the availability of the propagules. The fire weed (*Epilobium angustifolium*) with wind blown seeds has been known to colonize fire seed areas in Quebec with literally thousands of individual plants. The availability of suitable habitat and propagules under proper conditions for germination and establishment control the number, size and density of the population.

Change of Population Size and Density Over Time

Among trees and shrubs that live for many years younger plants get shaded out by older or faster growing individuals. The first seedlings of annuals in an area usually have a competition advantage for light, moisture and nutrients. This leads to the self thinning and the subsequent death of later annuals within the population.

Speciation occurs at Population Level

The actual process of evolution and speciation occurs at the population level rather than at the level of the individuals organism since the evolution is broadly defined is directional change in gene frequency through time, the changes in gene frequency only occur in populations of organisms further all population do not evolve at the same rate because different populations are under different selective pressures that effect, in different ways, the actual or potential evolutionary history of the population. Geographically isolated populations or groups of populations, which have only limited or no gene exchange with other populations of the same species. The similar population is normally such an isolate, the isolates are therefore particularly common near the periphery of the species range. Isolates are frequently sufficient difference to meet subspecies rank.

Interbreeding Population

A large inter breeding populations, containing many hundreds or thousands of individuals, would normally have a much larger gene pool than a small population of perhaps only a dozen or so individuals. A larger gene pool provides a **vital store house of genetic variation** which in turn provides a population with the capability of adapting over a period of time, to gradual changes in the selection forces of the environment. The gene pool of small population has proportionally fewer different alleles and consequently lacks the adaptive capabilities of the large population under changing environmental conditions. On the basis of chance a particular new gene combination is much more likely to become established in a small population than in large population. A large gene pool thus has a strong buffering effect that prevents rapid shifts in gene frequency and possible overspecialization.

The Chance Fluctuation of a Gene Frequency

The chance fluctuation of a gene frequency in a population is called *drift*. If the frequency of a particular gene drifts to either 100% or the gene frequency is fixed and since selection has presumably played no important role in the process it is called ***random fixation***. In small population random fixation is much more likely, more frequent, than in very large populations. Indeed it is very difficult, even with strong selection, to raise the frequency of a particular gene to 100% or to lower to zero in large population.

Maximum Evolutionary Potential

Population structure that provides maximum evolutionary potential is thus neither very large nor very small, but is considered to be a series of interbreeding populations of intermediate size. There can be rapid development of slight genetic differences as between some of the populations and then reblending of the slightly divergent gene pools. Such a reticulate breeding structure has the equivalent gene pool of a large population yet has much more adaptive flexibility without the disadvantages of the small populations.

Random Fixation : Founder Effect

At a new site in the establishment of a population, population size will most likely be small for the first few generations, thus regardless of the ultimate size of the population, the gene pool may show some of the effects of drift or **random fixation**, that were involved in the earlier stages of population establishment. Furthermore the entire fate of population may depend on the genes brought into the population when it was started. If, for example, a new population is started by a single seed that was transported to the new site, the very limited gene pool will produce a population of rather uniform individuals with a relatively low evolutionary potential. The reduction in population variability caused by the small size of the original gene pool provided by the organisms that started the population is called **founder effect.**

Population Size involves more than one Gene Pool

It involves some form of competition between individuals of the population, since population of organisms tend to occur in a relatively specific ecological situations which have finite limits, a second aspect of population size is population density. In plant population an increase in density means more direct competition between individuals for moisture, nutrients and light. The larger the population in a given area, the more severe the competition between its members and the more harsh the natural selection.

Population Size in Spatial Sense

In plants, population size in the spatial sense is established by certain environmental limits which define the ecologically specific soil type, type of rock outcrop, pollination range, or moisture condition may be only a few square meters in area. Population of plants wider physiological tolerance may occupy many square miles. Many populations of certain plants are more or less linear and follow stream banks, road sides, coastal dunes and similar linear habitats for considerable distances. However the population of a given species are never uniformly distributed with the geographic range of the species, are never individual plants usually, uniformly spaced within the population. Both populations and individual tend to occur in "clumps" or "patches" within their respective areas. The proximity of one population to another and the maintenance of a particular plant density within each population, becomes extremely important not only in the matter of pollination and seed production but also in regard to gene flow and abundant individuals and population variability which is of evolutionary and systematic importance.

Decrease of Tree and Shrubby Population

The trees and shrubs that live for many years and get larger each year, population density decreases as the population gets older and individual plants continue to die as they get shaded out by more robust or faster growing individuals in the population. This self thinning has also been observed in some annuals that maintain rather uniform density at maturity despite varied density of the seed at planting time. This comes about in two ways. In the first case, the first seedlings to germinate produce water soluble chemical substances that inhibits the germination of the other seeds in the immediate area. The first seedlings to stunt growth thus have a strong competitive advantage over the later seedlings that must compete with larger, established, seedling for light, water and walnut tree (*Juglans nigra*) are suspected of giving off substances which are toxic to many plants, including their own seedlings. In the second case, germination is not inhibited, dense stands of seedlings are produced and disease quickly removes many of the weak seedlings and population is lowered.

Role of Migration in Population Density

An individual plant obviously cannot migrate in response to increasing population density. Migration is still an important factor in relation to competition and density in plant population. There are numerous adaptations of fruits and seeds which promote migration away from the parent plant and often away from the population. For example the seeds of *Magnolia* have a bright red fleshly covering called an aril. When seeds are ripe, they hang exposed outside the follicles that make up the cone like fruit. The fleshy aril contains a water soluble chemical which prevents or inhibits seed germination only when the aril is removed then the seed will sprout. If a seed falls to the ground under the parent tree, it will not germinate. If a seed is eaten by bird, the flesh aril and its chemical inhibitor are removed as the seed passes through the digestive tract of the bird. A very hard seed coat prevents the digestion of the seed itself, therefore, when the bird excretes the seed, presumably at some distance away from the parent tree, it will be ready to germinate and the seedling will not be in fatal competition with the parent tree.

A high reproduction rate, if actually translated into high population density rather than into an expanded and becomes an adaptive liability that lowers the biological fitness of the organisms involved.

Outbreeding

Once organism becomes evolutionary adapted to a particular environment or set of selective forces, it can maintain itself indefinitely in a state of ecological equilibrium with other biotic elements to which it is adapted as long as the total environment itself remains essentially stable. Under such conditions, selection would be for continued stabilization of the organism increased uniformly rather than increased variability would have the most selective value, homozygosity would tend to increase and heterozygosity would tend to decrease. In plants, which have evolved a much more variable series of reproductive methods than have animals, strong stabilizing selection might be a shift to more inbreeding (self pollination; apomixes or asexual reproduction).

Among obligate outbreeding plants, it is possible that they might be highly heterozygous genetically even though they are quite uniform phenotypically. Also heterozygosity itself may increase individual survival values and also population fitness though heterosis and homeostatic effects, while the increased gene pool possible with heterozygosity provides a relatively high degree of evolutionary potential under changing environmental conditions or changes in the environmental selective forces that might be increased by migration.

Many of the evolutionary properties of populations depend upon the heterozygosity maintained by a relatively steady rate of recombination within the population. However in outbreeding populations chromosomal differences (inversions, translocations), assertive mating and occasionally, varying amounts of intrapopulation sterility provide a certain degree of restriction to complete or random recombination. Many flowers are bisexual or hermaphrodite, thus interbreeding at the level of the individual is possible for most higher plants. In these plants which are capable of self pollination are found a number of mechanisms that provide some recombination either through frequent or occasional out-crossing.

Cross Breeding has Survival and Evolutionary Values

Since interbreeding and outer crossing each has survival and evolutionary value under different conditions of natural selection. It is not surprising to find that many plants have evolved slightly ambivalent reproductive systems that allow for the effective operation of these divergent breeding mechanisms within a single population or in some cases a single individual.

In terms of inbreeding versus the degree of outbreeding three different systems can be observed:-

1. Predominant out crossing
2. Predominant selfing
3. Mixed selfing and out crossing.

   In addition, several highly successful methods of asexual reproduction have evolved among some of the higher plants and the combination of sexual and asexual reproduction adds to the evolutionary flexibility occurs of those plant groups in which it occurs.

6

Variation in Individuals and Populations

Objectives

1. To analyze and understand sexually breeding population and asexually reproducing organisms.
2. To illustrate the ecological differentiation of population and genotypic difference.
3. To have clear cut concept of phenotypic plasticity.

Variation in Individuals

A group of individual plants exhibit differences between the genetic architecture of plants. No two individual plants on the face of the earth exactly look alike. Individuals of the same population of sunflower exhibit a lot of variation.

Individual plants exhibits individual variation in terms of plant height, nodal length and arrangement of leaves, leaf numbers, leaf length, leaf breadth, flowering times since variation is law of life.

A group of individuals of same species, sharing same habitat or discrete units' places and reproduce and complete their cycle for that period are called population.

Definition of Breeding Population

A breeding population can be defined as a "reproductive community of sexual and cross fertilized individuals that share in a common gene pool" (Dozhansky, 1950). This kind of population, called a breeding population or sometimes a Mendelian population has to be statistian or the demographist who also define a population as a group of individuals but with a different binding property such as "interaction in space" or inhabitants of a given geographical locality.

Sexually Breeding Population

Breeding populations are not necessarily discrete units often it is hard to draw clear separating lines between them. A good example is the belt of balsam fir- *Abies balsamea,* that extends across the northern Canada from New foundland to Alberta and from New England to Iowa. Sometimes for hundreds of miles with no obvious interception. How can populations be demarcated in this case. Does the impossibility of delimiting the populations invalidate the concept.

The inability to differentiate distinct gene pools in certain cases and the overlapping of gene pools in others does not necessarily mean that they donot exist. The problem of delimiting the breeding population into discrete units may be compared to the problem of delimiting young and old organisms. Neverthless young and old are useful and valid distinctions.

Asexually Reproducing Organisms

In asexually reproducing organisms there are no breeding populations, because each individual can reproduce independently. However, *most asexually reproducing plants occasionally revert to sexual reproduction.* Formation of seeds without recourse to fertilization a process known as apomixis. However, occasionally some seeds are formed in the normal sexual way by way of union of two gametes an egg and sperm. A population of asexually reproducing plants is a breeding population to the extent that it reproduces sexually. In a population of self-fertilizing plants some plants normally out cross, the concept of breeding population is usually applied also to self fertilizing organisms.

Aggregation of Similar Plants

Physical proximity or closeness or aggregation of similar plants is not sufficient to constitute a population, there must be interbreeding. All the plants in population are therefore potential mates. The genes of the parents are combined in the offspring so that latter usually have intermediate characteristics. Some combinations of genes cause some individuals to have a greater fitness than others. Over the course of time, the random, crossing of the individuals in each generation will produce many kinds of phenotypes, and those phenotypes that have the greatest fitness will tend to become more numerous and those that are least fit will disappear. As a result of the combined action of natural selection and the sexual process provided that the environment is uniform over the area, occupied by the population, the individuals of the population should tend to have similar genes and also should tend to look quite similar.

Ecological Differentiation of Population

Massarat, conducted experiment on *Polygonum amphibium* and arranged to cultivate on dry sand, moist soil, and in water produced three distinct phenotypes. From his experimental results, Massart concluded that corresponding forms in nature had arisen in a similar manner through direct modification by environment, it would provide an example of the morphological differentiation of populations by habitat modification of genetically uniform original stock. Swedish Turesson pointed out in 1922 at the beginning of his own studies on the differentiation of plant populations, Massart's experiments prove nothing of the sort. The three forms of *Polygonum amphibium* found in nature may be no more than modifications of similar genotypes or they may possess dissimilar genotypes, differing from each other genes concerned with physiological and morphological adaptation.

From the mere inspection of phenotypes there is no way of determining between true genetical difference and simple environmental modification.

Establishment of Genotypical Difference

One way of proving Massart's contention is to have cultivated plants of the three forms, taken from nature, side by side in a common neutral habitat. Morphological convergence would then have been strong evidence by a demonstration that all three forms could be inter converted by reciprocal transplanting, failure of the transplants to converge in a neutral habitat would suggest that the forms might be genetically dissimilar and a demonstration that the progeny of the three were not alike even when grown from seed in the same habitat would provide further evidence of genotypical difference could be finally established by crossing experiments.

Experimental cultivation is thus a technique of importance not only in exploring the phenotypic plasticity of single genotypes, but in detecting differences between genotypes. The essence of this was appreciated by deCandolle, who advocated the cultivation of critical forms side by side in various soils to see whether or not their differences would eventually disappear.

Jordan Experiments on Hereditary and Non Hereditary Forms

A Jordan used this method of comparative culture to distinguish between hereditary and non hereditary forms in great many species of French flora. Jordan sought to demonstrate that the Linnaean species were often large aggregates which could be broken up into minor units which perfectly limited and distinct, constant and invariable in their differences which he took to represent the real species. As might be expected, what Jordan often succeeded in defining were individual biotypes possessing distinctive characters owing their perpetuations to reproductive peculiarities. Thus the spring whitlow grass *Erophila verna* was pulverized into more than 200 species, many of those are now known to be obligately self pollinated pure lines. In outbreeding species Jordan was less successful in defining permanent or even persistent forms many of those he described being simply biotypes with distinctive character combinations, sorted out fleetingly from the flux of forms occurring in wild populations.

Phenotypic Plasticity

The phenotypic variability(plasticity) of plants probably has a great survival value, because it is such a universal phenomenon. It is a phenomenon that is largely peculiar to plants, because animals, particularly the higher vertebrates maintain their phenotypic characteristics much more constantly than plants when put in different environment, plants cannot move from place to place and cannot avoid the environment rigors of the place where they happen to germinate would show the greatest degree of phenotypic variability. Even the animals such as squirrels cannot avoid such as low temperatures and meager resources in winter by a somewhat similar change in phenotype e.g., accumulating winter fats acquiring long fur and hibernating.

Tridax Procumbens Population

In India Tirdax procumbens population grow very well luxuriously in the rainy season with several leaves reaching the large size of 26 to 36 cms or so. In receding monsoon the surviving plants produce small sized leaves and stems to minimize the transpo- evaporation. Thus the plant population which have been surviving in winter parsimoniously spend water and tide over the adverse climatical conditions such that tht large sized leaves in rainy season and very small sized leaves in the winter season. The possession of different types of different sized leaves are adaptations to the existing environment in favorable and adverse climatical conditions respectively. The response of different features of the plant to different environmental conditions or factors varies considerably.

Unchanged Characters - Good Characters

Some characters such as leaf arrangement or floral structures remain more or less unchanged under a range of different conditions. These are frequently referred to in taxonomy as 'good' characters because of their relative constancy and are regarded as 'genetically fixed'. Other features such as leaf shape, stem height, time of flowering etc., develop differently under different environmental conditions. ***The term phenotypic plasticity is applied to these variations shown by the phenotype in response to environmental fluctuations.*** The same genotype may, according to the environmental factors prevailing, give rise to different phenotypes. This plasticity may be so wide as to overlie or even obscure the genotypic variability. In fact, the distinction between genetically determined characters and environmentally induced characters is a false one since both types of character are the result of genetical inheritance and environmental control. In other words, each genotype has a range of plasticity which is genetically determined. In some cases the range of expression is narrow the plants are termed stenoplastic or broad, being called euryplastic. It is the inherited response which produces the character or phenotype not just the environmental factor. The adaptive or non adaptive nature of these differences can be understood in connection with ecotypes.

7

GEOGRAPHIC VARIABILITY

Objectives

1. To study the variability of organisms in different geographical locations.
2. To check the characters of phenotype are a reflection of the interaction of the internal genetic factors of the plant with environment.
3. To assess the correlation that existed among phenotype, genotype and environment.

Introduction

The basis for the recognition ecotypes within a species is simply that the occurrence of population with certain heritable adaptative characters in common should be correlatable with particular conditions of the habitat. Although much intraspecific variation of this nature, it is commonly found that populations from different areas differ morphologically even when no obvious differences in their habitat exists. Furthermore it is often the case that over and above the local ecological variation, there is regional variation which can hardly be looked upon as directly adaptations.

Geographical Distribution

The transverse lines on the globe are called latitudes. Different latitudes are located differently and possessed different climates. The plants of some species living in different distantly placed latitudes exhibited different morphological characters which are associated with habitat. *Gutierrezia sarothrae* a globose shrubby plant belongs to the compositae grows from western edge of the prairies in North America to the pacific ocean and from northern Mexico to Alberta and British Columbia in Canada as shown in figure 7.1

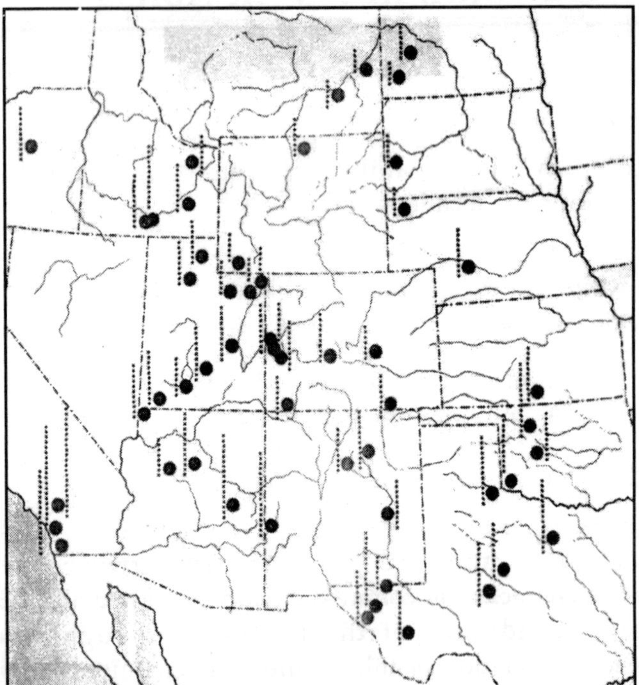

Fig. 7.1: Average plant height of populations of *Gutairrezia sarothrae*. Each division of the bar represents one inch. Note how northern populations are generally shorter than southern ones, and how western populations on the same latitude are generally larger than the eastern ones.

Population Size

Solbrig, (1964) found very discrete, widely distributed scattered populations which formed by the fifty to one thousand plants or more. These populations are separated from each other by anywhere between a few hundred yards to tens and even hundred miles.

Sample Size of Population

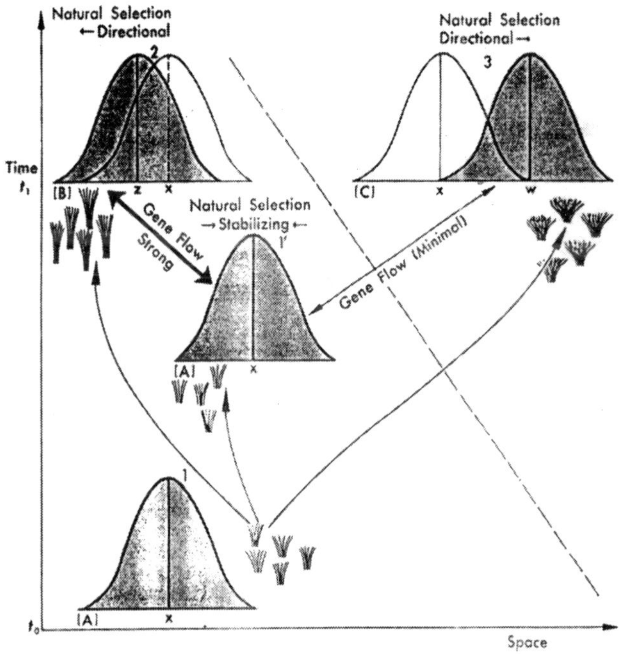

Fig. 7.2: This pictorial graph illustrates some of the evolutionary forces at work in a situation involving geographical variation in space and change in time. At time t_0, population 1 was adapted and growing at site A with a mean x (to represent a mean of all character states). At time t_1, population 1 has given rise to population 1′, 2 and 3 at sites A, B and C. At B natural selection favours a mean z, at A it favours still a mean x and at C mean w. A and B being very close and environmentally similar, gene flow is strong, and directional selection not very different. The opposite is true for C.

In order to obtain the degree of the patterns of variability in these populations (eleven morphological characters were measured) as shown in table in each of fifty plants from every one of fifty three populations differed from each other in the mean value of the measured characters in a statistical way). The most striking example is offered by the number of flowers in the inflorescence, called a capitulum in compositae. In samples of fifty plants from single population the number of ligulate flowers varied from two to seven and the number of tubular flowers from zero to seven. Traditionally this character has been considered very important in the characterization of species in the family compositae.

Correlation of Characters with Geography

The value of the mean for some of the characters measured was correlated with geography e.g., plant height (fig-7.1) is correlated with latitude other factors being equal, northern plants are shorter than southern area and longitude: other factors being equal, plants growing west are larger than plants growing east. Factors other than distribution in space are also correlated with plant height. Within the same latitudes, plants of *Gutierrezia sarothrae* of higher altitudes are shorter than those of lower altitudes. The number of flowers per head on the other hand were not correlated with geography if correlation existed it was not absolute. As shown in table 7.0.

The pattern of variability that is found in *Gutierrezia sarothrae* to a greater or lesser extent in most species that are distributed over a large and diverse ecologically areas. This is pattern of correlation of the variability with geography for some characters, but not all.

Individual populations are subjected to environments that vary from population to population; the phenotypes that are favored by natural selection at one site are not necessarily favored at another site. The evidences deduced from transplant experiments indicate that natural selection will favor different phenotypes in different areas. On each site certain characteristics of the phenotypes are of more direct adaptive value than others.

Table 7.0 : Comparison of nine characters of Diploid and Tetraploid Plants of *Gutierrezia sarothrae*. [From Solbrig, 1964.]

Character Examined	RANGE Diploid	RANGE Tetraploid	MEAN Diploid	MEAN Tetraploid	t* [P0.05-1.96]
Plant height	4-38	4.5-28	10.93	10.13	0.942
Involucre height	2.5-5.0	2.6-6.2	3.77	3.56	0.075
Involucre width	0.9-2.9	1.1-2.7	1.77	1.81	0.031
Tubular flowers	0-7	0-7	3.67	3.90	1.118
Ligulate flowers	2-7	3-6	4.87	4.49	0.851
Tubular pappus Length	0.5-2.0	0.6-2.0	1.24	1.31	0.076
Ligulate pappus Length	0.2-1.2	0.4-1.3	0.58	0.73	0.333
Pollen diameter	9-16	11-17	12.57	13.59	2.930
Stoma length	14-24	15-25	18.13	19.97	1.870

*A value for t less than 1.96 means that the difference between the diploid and tetraploid plants is not statistically significant at the 5% level, or in other words, the probability of error in considering the populations different, is more than 5%.

These more important characteristics are the ones that will show the strongest correlation with geography. However, the transmission of pollen and seeds from one population to another will introduce genes from neighbouring populations. These introduced genes will tend to be eliminated in the new host population unless they contribute to its total fitness. If the rate at which they are eliminated is slower than the rate at which they are introduced gene flow will prevent the populations from diverging significantly. Gene flow in this case acts as a cohesive force within the species.

Spatial Selection

On the other hand, if gene flow becomes negligible or is completely eliminated (because of the distance between populations, for example), the rate at which natural selection proceeds is accelerated. Spatial isolation has also been considered as a first necessary step in the speciation process as we will see further on.

Variable which cannot be Tested

The hypothesis of geographical variability is simple; however, the natural selection is highly involved, and the demonstration of the hypothesis is quite difficult in specific cases. The basic variables, time, space, gene flow and intensity of selection, are not correlated. Furthermore, in most cases only space can be measured among the variables. Time since establishment of the population is almost impossible to determine in most cases, and estimates of gene flow and intensity of selection have so far been very hard to obtain. Consequently, the biosystematist is usually faced with the following situation. A careful analysis via a statistical description of the phenotypes of the population (morphology, anatomy, chemistry, and so on) establishes that there is variation. But unable to determine the intensity of selection or the degree of gene flow, the biosystematist analyzes the geographical picture and the broad features of the environment looking for correlations. These trends are then interpreted in terms of selection, as in the example of *Gutierrezia sarothrae*.

Test the Inheritance of the Characters

However, this general approach is not sufficient for an understanding of the dynamic picture. Measuring plants in the field, no matter how carefully, only reveals the phenotypic differences, and these are due to an interaction between genotypes and environment. A necessary step is to determine the extent to which the differences are heritable. This determination is accomplished by growing plants from various geographical sources under a uniform environment, preferably under several such uniform environments. Phenotypic differences can now be assumed to represent genetic differences. Crosses between the populations can then be performed to test the inheritance of the characters under study. Furthermore, the performance of the hybrid can be used as an estimate of the effect of gene flow (e.g., are the hybrids weak or vigorous?), as well as the degree to which populations have differentiated in regard to genetic isolation barriers (how much seed does the hybrid produce?). Finally, a careful analysis of the local environments should reveal more features to which the local population may be closely adapted. Such studies are time consuming and require special facilities, such as transplant stations, growth chambers, and so on, that are not always easily available. When these studies have been performed they have revealed that the differences between populations of the same species, particularly physiological differences, are always of a greater magnitude than would be expected from a cursory analysis.

Turesson's Pioneer's Experiment

Actually northern Europe is in northern altitudes and with higher latitudes. Whereas southern Europe is situated in different latitudes with different climatical conditions. The Swedish botanist Gote Turesson (1922,1925,1929) conducted experiments and he assembled plants from diverse environments of Europe at the experimental garden of the Institute of Genetics at Akarp, Sweden.

Physiological characters such as time of flowering were especially emphasized. He found that, in general, southern European, Alpine, and high Nordic populations were first to bloom at Akarp, whereas lowland forms from northern and central Europe were the last to bloom. He was also able to demonstrate that certain morphological types tend to be associated with particular habitats. For example, Alpine forms in general were dwarf and low land types were tall, whereas plants growing at the **seashore** were often characterized by a prostrate habit with thick, fleshy leaves. Forms from southern latitudes were, as a rule, taller than plants from the far north which resembled the Alpine forms in height.

As a result of his studies, Turesson proposed the concept of ecological race or ecotype, also known as genetical race or simply race. The basic concept suggests that at a given locality the local populations possess particular heritable physiological characteristics that make them better adapted to the local environment than populations from other localities and that species are composed of different races.

8

Transplant Experiments

Objectives

1. To understand how distinct phenotypes are formed in different habitat.
2. To understand how the distinct phenotypes which are formed in different habitat - exhibits the same in the uniform habitat.
3. To assess the performance of different phenotypes in different habitats of experimental stations.
4. To group the generations about the transplant experiment.

Introduction

In the varied environment transplant experiments, plants from the different ecotypes were cloned and ramets were grown at the different stations along the altitudinal transect. The outstanding result was to reveal that each possessed different capacities for response, both physiological and morphological and that in most cases their individualities were retained at the different stations. In general morphological changes induced were slight in comparison with the far reaching changing in the annual cycles of development which plants of the various ecotypes were underwent when taken outside their natural ranges.

Turesson Experiment

Swedish botanist Gote Turesson (1922, 1925, 1929) was one of first man to apply the transplant method of study to geographical variations. He collected plants from diverse environments and grown there in the experimental garden of the Institute of Genetics at Akarp, Sweden. Species which are growing in continuous geographical distribution over a variety of climatic physiological races are twenty or more than twenty. By comparing such collections in a standard environment, Turesson could able to get an insight into the phenomenon of geographical variation.

The flowering time and plant height were considered as standard characteristic features. He observed Southern European, Alpine and higher Nordie populations were first to bloom at Akarp where as low land forms from northern and central Europe were the last to bloom. He also observed that alpine forms showed short stature whereas low land types were tall and while plants growing by the side of sea shore were shown prostrate habit and fleshy leaves. Forms from southern latitudes were as a rule taller than plants from the far north which resembled the Alpine form in height. From his results Turesson proposed the concept of ecological race or ecotype, also known as genetical race or simply race. In the given locality the local populations possess particular heritable physiological characteristics that make them better adapted to the local environment than population from other localities and that species are composed of different races.

Transplant Experiments of Clausen, Keck and Hiesey

Clausen, Keck and Hiesey conducted transplant experiments for nearly 30 years. Their investigations were established under the auspicious of the Carnegie Institution of Washington at three experimental stations in California at 38 N latitude. The first is located at Stanford near San Fransisco Bay. It has a mild coastal climate, the altitude of station is 300 mts although mild winter frosts do occur as shown in figure 8.2. In general the conditions are favourable for a considerable amount of growth even during the winter because the minimum temperature remains relatively high, winter conditions are much less likely to hinder the growth at Stanford than are the heat and drought of the summer.

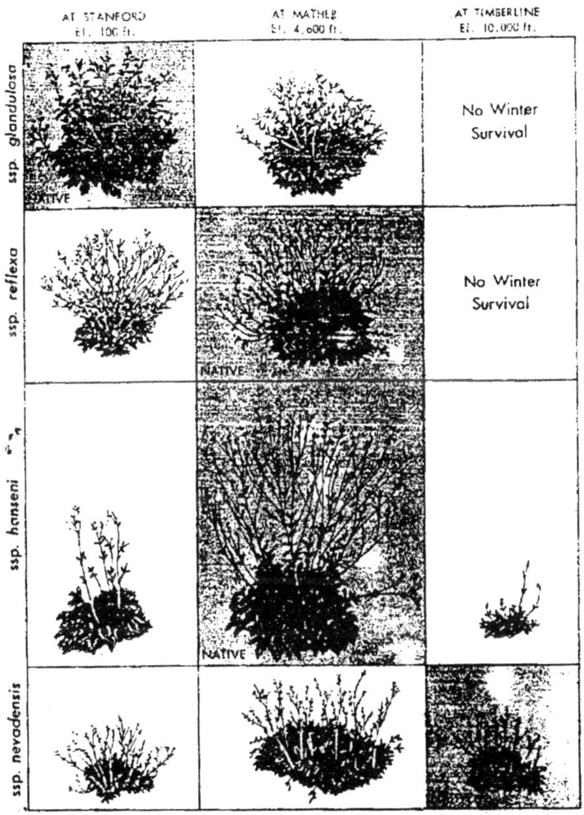

Fig. 8.1. Responses of four ecotypic subspecies of *Potentilla glandulosa* to the three contrasting environments in the Carnegic Institutions gardens at Stanford (left), Mather(centre) and Timberline (right). The four subspecies are : *necadensis* from 10,000 feet elevation (bottom row), *hanseni* (second row from bottom); *reflexa* (third row from bottom), from 2,500 feet in the Sierra Nevada foothills; and *grandulosa* (top row) from 900 feet in the Outer Coast Ranges of california.

The second station is located at Mather, at 1400 m elevation on the west slope of the Sierra Nevada. In Mather the climate favours the growth of the plants only during the period between May and October. Spring weather starts in the first week of April but frost may occur until the beginning of the June. The summers are mild with warm days and cool nights and little rain. The fall is cool and dry. Snow can be expected between October and April rarely exceeding 3 feet in depth. The third station is situated at Timberline in the Sierra Neveda at 3050 m. The climate at Timberline is very severe with only three to six weeks frost free weather during year and frost may occur throughout the year. Snow will accumulate to 20 feet in the winter.

Clausen, Keck and Hiesey (1940,1946,1948) and Clausen and Hiesey (1956) observed the existence of local races in certain species like *Yarrow, Achillea* and species of subtribe Madineae of the compsitae. The plants of *Potentilla glandulosa* were brought from the fields divided into clones and one planted at each of the three station for all those four sub species *P.glandulosa, typica reflexa, hansei and nevadensis*. The observations of Turesson were confirmed. In the observation it was found that plants from low land did not survive at Timberline and plants from the high altitudes likewise grow poorly at low elevation. This was time even among population of the same species. Crosses between these different populations showed that differences were genetically determined and that they constituted ecotype in same sense of Turesson's definition.

Potentialla Glandulosa Species

Potentialla glandulosa herbaceous plant belongs to the family Rosaceae. It is formed a series of populations scattered widely through western northern America. A great number of populations from California were sampled by the team of Clausen, Keck, Hiesey. The species can be divided on the basis of morphology into four subspecies. Subspecies *glandulosa typica* occupies the coast range of California end isolated localities in the foot hills of the southern Sierra Nevada.

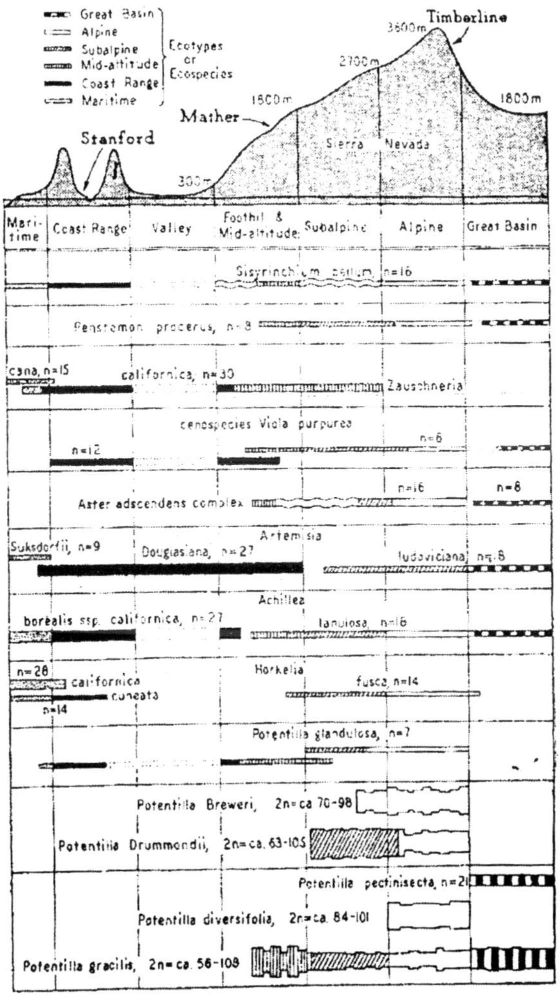

Fig. 8.2. Distribution of major climatic races with species complexes in a transect across central California. The width of the bar symbols roughly indicates the degree of polyploidy. (From Clausen, Keck & Hiesey, 1940.)

Subspecies reflexa occurs on warm, sunny slopes of the foot hills and mid altitudes of the Sierra Nevada from 300 to 2,000 m elevation. Subspecies *hanseni* is a meadow form occurring between 1300m and 3000m altitude and finally subspecies *nevadensis* is a high altitude meadow or slope form that occurs 3,000 to 3600 m altitude.

The subspecies of *Potentialla glandulosa* have no barriers to interbreeding except through spatial and ecological separation.

Sample from several populations from each of these subspecies were grown at the three transplant stations in Stanford, Mather and at Timberline with following results-

1. The coast range form of subspecies *glandulosa typica* grew taller at Stanford less well at Mather and failed to survive that winter at timberline
2. Subspecies *reflexa* on the other hand grew best at Mather, where it is native less vigorously at Stanford and also died at timberline.
3. Subspecies *hanseni* was strongly winter dormant at Stanford grew best at Mather and survived at Timberline it was unable to set ripe seed at the high altitude.
4. Subspecies *nevadensis* barely survived at Stanford where it was winter dormant. It was native to Mather and it was vigorous and consistently set seed at Timberline.

These four morphological subspecies also represent four distinct physiological types. Each one was performed best in terms of growth and seed set at the station that was closest to its native environment with exception of subspecies and *nevadensis* which did better at Mather than at Timberline. Subspecies *nevadensis* was subspecies capable of functioning normally in the extremely harsh conditions of the timberline station.

Macmillan Transplant Experiment

Macmillan (1959, 1964, 1965) studied the nature of the grassland communities of North America. He analysed the behaviour of hundreds of clones of four species of grasses namely *Andropogon scoparius, A. gerardi, Panicum virgatum* and

Sorghastrum nutans. These populations occurs as far north as Michigan and New England and as far south as Mexico city were sampled and grown under uniform conditions in the experimental garden at the University of Texas at Austin, Texas. Macmillan observed hereditary morphological and physiological variations between clones from different geographical regions. These did not form well delimited groups but rather tended to integrate slowly into each other. These variations are clearly adaptive.

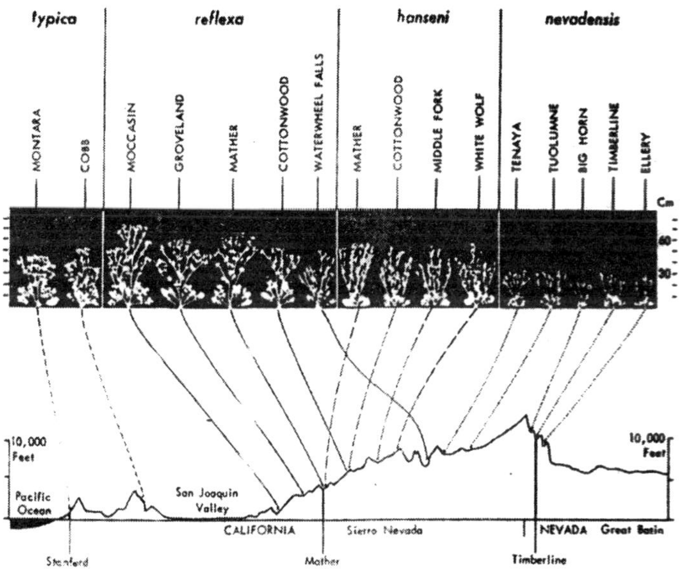

Fig. 8.3. Silhouettes of herbarium specimens of sample plants of *Potentilla glandulosa* of the four subspecies grown at Stanford under uniform conditions, with a profile across central California at approximately 38^0 north latitude, indicating the original localities of the populations. Note how the morphology of the plants becomes modified from east to west, and also note the ssp. *Reflexa* and ssp. *Hanseni* overlap in part of their range. The profile map also indicates the position of the three transplant stations at Stanford, Mather, and Timberline. (Modified from Clausen, Keck and Hiesey, 1940, by permission.)

Different Species Growing in Northern and Southern Regions

For example: The plant species from northern clones (Massachusetts, Michigan, Wisconsin) were of shorter stature and were early flowering whereas the plant species of southern clones (Texas Louisiana) were taller and late blooming. It was hypothesized that the climatic pattern of northern habitats with relatively few frost-free periods of growth, selection favours a developmental pattern geared to maintaining the populations through rapid growth and early flowering.

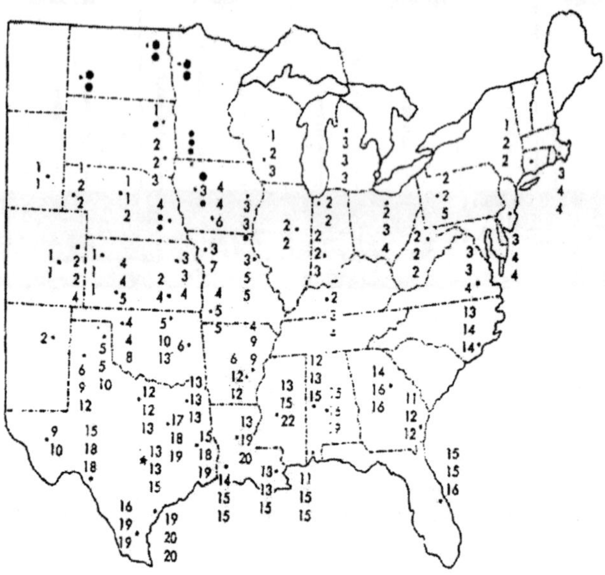

Fig. 8.4. Flowering of transplanted clones of *Panicum virgatum* at Austin, Texas (indicated by a star). Numbers indicate weeks after April 27, 1961, for the first frant to bloom in each clone. Large circles indicate that the plants from those localities never bloomed. (From Macmillan, 1965, by permission.)

In southern climates, on the other hand, the growing season is much longer. In this situation the interpretation of the greater stature of southern grasses is that there is a competitive advantage for the plant, that can grow taller and capture more of the Sun's energy. Probably there is also an advantage in shading (and killing) smaller plants.

During the long growing season of southern regions, plants probably receive selective advantage by continous vegetative growth and late blooming. Plants in northern communities with such growth pattern are eliminated by frost.

Ecological Races

Fig. 8.5. Height of plant (culm) at time of flowering, for transplanted clones of *Panicum virgatum*, grown in Austin, Texas, during 1961. No. 1: 1-25 cm; 2: 26-50 cm; 3: 51-75 cm; 4: 76-100 cm; 5: 101-125 cm; 6: 126-150 cm; 7: 151-175 cm; 8: 176-200 cm; 9: 201-225 cm; and 10: over 225 cm. Large circles indicate that the plants of those localities did not bloom. Note the great difference in size between northern and southern plants. (From macmillan, 1965, by permission.)

The concept of the very distinct ecological race is not as easily applicable to prairie grasses as it is to certain California plants that grow in very distinct ecological regions that donot intergrade. However the basic concept that species are formed by a series of more or less locally adapted populations seems to be confirmed. The answer to the problem of how these local races can originate in the presence gene flow from other populations has two alternative solution. One hypothesis is that there must be very little gene flow when the local races are being formed. The second hypothesis that selection is sufficiently intense so that gene flow is not an important consideration. All these hypothesis are not mutually exclusive, and all combinations of degree of gene flow and intensity of selection can be visualized as leading to the formation of local races.

In some cases such as California the geographic and physiognomic situation is such that it creates disfunction of given a species the central valley of California species the Coastal race, subspecies *glandulose* of Potentilla glandulosa from the foot hill races, subspecies *reflexa*. In many other cases sharp climatic boundaries exist that are not so obvious as the geographic ones, but are equally effective in checking gene flow. The populations in different climatic regimes tend to produce the heritable physiological morphological variations with which they will survive in those habitats and forms ecotypes of subspecies.

Watson and Clausen (1961) Transplant Experiments

Watson and Clausen conducted experiments on the grass genus *Poa*. Many species of this genus are apomictic that is, they reproduce seeds that are asexually. However, these apomictic are facultative, meaning that some seeds can be produced by sexual means in each generation. Taking advantage of this fact, a cross between two related species *Poa ampla* and *Poa protensis* was produced. From the offspring of this cross(which reproduced entirely by sexual means) several apomictic lines were selected. The offsprings of each of these lines was genetically identical or almost identical, because they were produced without recourse to fertilization. These were then grown and studied at three distantly located places after dividing three equal sets of seeds.

Different Experimental Stations

The first experimental station was the experimental garden of the Carnegie institution of Washington at Stanford, California located at 38 N, USA.

Table 8.0: Comparison of poa Hybrid and Apomictic Lines Grown at three different localities. [From Watson and Clausen, 1961.]

Center	Character	Parents		Lines				Center
		pa	pp	S–23	S–309	P–507	E–701	Average
PENTLANDFIELD 1957-58	Stem length	100.0	81.7	77.4	84.9	93.4	85.2	87.1
	Stem number	115.4	107.7	93.7	100.3	50.9	116.4	97.4
	Basal width	11.1	55.2	45.8	48.4	40.9	40.8	40.4
	Vigor	7.6	7.6	8.4	7.5	6.6	6.7	6.1
	Flowering date	12/VI	23/V	10/VI	30/V	3/VI	21/VI	7/VI
PULLMAN 1956-57	Stem length	83.7	86.7	75.0	83.3	104.1	73.7	84.4
	Stem number	153.3	391.0	207.0	316.5	369.1	260.5	289.9
	Basal width	13.4	44.8	27.5	25.8	34.2	28.5	29.0
	Vigor	5.9	7.7	5.5	6.9	8.2	6.2	6.7
	Flowering date	31/V	6/V	1/VI	26/V	20/V	28/V	18/V
STANFORD*	Stem length	69.2**	76.8	58.8	81.9	82.8	38.8	68.1
	Stem number	69.2**	132.1**	53.3	61.5	99.0	30.0	72.5
	Basal width	12.1**	34.3	22.7	23.6	27.6	38.5	26.5
	Vigor	4.2**	5.2	5.3	6.8	6.9	4.9	5.6
	Flowering date	23/V**	13/IV	21/IV	15/IV	16/IV	12/IV	22/IV

* Values taken in different years between 1950 and 1958.

** Not many plants survived for measurement.

The second station was the soil conservation service, Nursery at Pullman, Washington, USA situated at 46 N. The altitudes here is 300m and the winters are moderately severe, including some very low temperature. The third station was the experimental garden of the Scottish Plant Breeding station at Pentland field near Edinburgh, Scotland at 56 N. The altitude here is 200m and growth is favored during the spring, summer and autumn only as the average winter temperatures being lower than at Stanford in spite of relatively mild winter temperatures. As shown in table 8.0

Different Habits Such as Tall Growing, Short Stature Plants

The parental species have very different characters. *Poa ampla* is rather variable species nature to the pacific northwest of North America. It is very tall growing, winter active bunch grass. *Poa pratensis* have a very wide distribution, through out the northern hemisphere. It is not robust as *Poa ampla*. Table shows the mean of four characters measured and the date of flowering at the three stations for the parental species and four apomictic hybrid derivates. Several important points are demonstrated. First, identical or almost identical genotypes produce dissimilar phenotypes when grown under different conditions. Second, related but different genotypes behave differently under the same environmental conditions as seen by the four lines selsected from the F2 of *Poa ampla* x *Poa pratensis* cross. Finally different characters are affected in different ways by the environment. For example, in the line E.701 the number of stems was considerably larger when the plants were grown at Pullman (260.5) than when grown at Penltand field (116.4). At Stanford the number was considerably smaller (30.0). Although the stem length was about the same in Pentland field (85.2) and in Pullman (73.7) but again much smaller at Stanford.

9

SPECIATION

Objectives

a. To understand the conceptual differences between evolution and speciation

b. To understand the real correlation between the morphological variation vs reproductive isolation, morphological similarity vs reproductive isolation, morphological similarity vs genetic isolation.

c. To understand dynamics of evolution

d. To understand and assess the different types of species

Introduction

Biological evolution is usually considered a series of processes involving descent of organisms with modification marked by successive adaptations to environmental conditions, governed by competition and natural selection acting on variation. The general purpose of evolutionary study is to determine the origin of new taxa and to interpret the historical development of the old. The fundamental components of evolution are variation, speciation, and phylogeny with basic activity of analysis of character variations for 1) the generation of character correlations for organisms, populations and species 2) the determination of the origin of new taxa and 3) the interpretation of historical development (origin, maintenance, trends, pathways of the old). Existing species are derived from the preexisting species with modification in geographical regions (allopatric). Biosystematist as student of speciation select group of organisms for analysis of character variation, determine the casual mechanisms for the variation and variants within the population and ascertain the isolating mechanism controlling interbreeding between populations.

In biosystematics, one of the most important intricate problems is the definition and determination of species. The two facets of the species problem involve the differences. The first facet of these is the delimitation of the evolutionary unit. The second facet is the taxonomic problem of concretely identifying specific set of individuals, as belonging to a particular species. In the biological point of view the problem can be considered specifically on process that lead to the formation of species. Second aspect is on how species can be defined for taxonomic purposes.

The Concept of the Species

The problem of the nature of basic unit that was being classified, namely, the species. The broad application of useful and workable system of nomenclature and a system of classification during the last century brought into focus.

Different Definitions of Species

In logical terms the species is a class to which belong all the objects that have certain common properties.

Morphological species: Species are characteristics of the phenotype usually, but not necessarily always, pertaining to the exterior morphology.

What are the properties that define species and what are the biological process underlying their existence. The properties used to define a species characteristics of the phenotype usually not always by a single character. However as species is not defined by only one character, there must be a series of correlated characters. This way of dealing the species has been called **morphological species concept**. The number and kind of characters by which a morphological species is defined varies widely according to the group of plants being classified. But it is undesirable fact that most of plants can be grouped into species on the basis of a series of correlated morphological characters alone.

Attributes of a Species

a. Species is a genetic unit consisting of large, intercommunicating gene pool.

b. Species occupies a specific ecological niche, unoccupied or unutilized by another closely related species.

c. A species is an ecological unit

d. Species interacts with the environment and other species present in a given environment.

e. Allied species, though occupy separate areas of distribution but these are often adjoining.

f. Species have fully effective reproductive isolating mechanisms.

g. Intermediate or transitional forms are usually absent.

Evolutionary theory brought into focus, the genetic relationships between members of a breeding population as well as their gene continuity through time. This evidently led to a redefinition of the species in genetic terms. A redefinition that has been called the biological species concept first introduced by Dobzhansky (1937) and latter redefined and amplified by Mayr (1942, 1947, 1957, 1963).

Species are groups of actually or potentially interbreeding natural populations, which are reproductively isolated from other such groups, according to new biological species concept, species are classified by only one criteria: reproductive isolation.

From the redefinition of the species the 1st question emerges is the correlation between the species defined in terms of morphology and in terms of reproductive isolation.

Correlation Establishment Between Morphology and Reproductive Isolation

The species belonging to different families, most of them are not related and they are morphologically distinct and reproductively isolated. This is exemplified by the genus *Glandularia.*

The genus often also considered as section *Glandularia* of the genus *Verbena,* and found in South America where it grows from Peru to Argentina, Guatemala to the Central and Eastern United states in North America. Species of *Glandularia* are easily distinguishable from the species of the closely related genus *Verbena* and repeatedly. When the crosses between the species of the two genera were attempted, no seed was produced. However crosses between most of the species of Glandularia can be produced.

Let us examine the three South American species *G. Peruviana, G. Pulchella* and *G. Santiaguinsis.* These three species occupy approximately same habitat, open, dry and semi dry grassland in northern and central Argentina. *Glandularia pulchella* and *Glandularia santiaguensis* replace each other geographically, with the former growing in the cooler southern and eastern areas and the latter in the more subtropical northern and western areas. Whereas *Glandularia peruviana* stretches all over the territory occupied by the other two species. In *G. Peruviana* the leaves are entire,more oppressed habit, larger internodes, large red flowers, lack of glandular appendages and the species differs with other species. Glandularia and flowers and less divided leaves as shown in figure 9.1. Hybrids among the three species are rare and occasionally seen. When artificial crosses are made in experimental garden they exhibit 40-50% pollen fertility, and was observed only about 1% of the flowers set seeds. The genetic isolating barriers among the three species is about the same and that species almost reproductively isolated.

G. megapotamica and *G. moricolor* on the otherhand yielded hybrids that were more fertile. The crosses between *G. peruviana* and *G. megapotamica* for example was approximately 65% pollen fertile and the cross between *G. peruviana* and *G.*

moricolor was approximately 80% pollen fertile. Morphologically *G. megapotamica* is similar to *G. peruviana* although it has flowers of a different colour and is more erect. *Glandularia moricolor* on the other hand is quite distinct, having elliptic leaves, erect habit, smaller and deep purple flowers, and a more compact inflorescence. Neither of these two species grow together with *G. Peruviana*. *Glandularia megapotamica* grows in subtropical gallery forest of northeastern Argentina, Brazil and Paraguay whereas *G. moricolor* is a species of the margins of the subtropical forests of northern Argentina. Consequently, the genetic isolation, between these two species and *G. peruviana* is not complete. However, they are effectively isolated reproductively because they do not grow together.

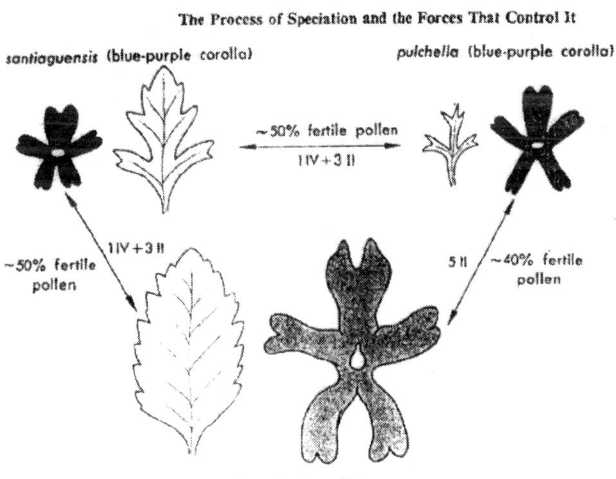

Fig. 9.1. Leaf shape and corolla of three species of *Glandularia*, and the pollen fertility and chromosome pairing relationships of the hybrids among them. Note that although *G. pulchella* is morphologically less similar to *G. santiaguensis* and *G. pulchella* than these two are to each other, the hybrids between all three species are approximately equal in pollen fertility. As to chromosome architecture, *G. santiaguesis* is more dissimilar to the other two species than they are to each other. These three species illustrate the reticulate nature of evolution.

Correlation Between Morphological Differences and Reproduction Isolation is Universal or not Lack of a Reproductive Isolation is Found Among Many Species of Oaks Other Kinds of Trees

C.H- Muller (1952) examined four species of oaks in Texas. He found that they hybridize readily and that apparently their hybrids are partially fertile. Nevertheless each of the four species is adapted to a different type of soil 1) *Quercus mohriana* to limestone, *Q. havardi* to sand, *Q. grisea* to ignaceous outcrops, *Q. stellata* to clay. Hybrids are able to grow only in areas where the various soil types overlap. Since these hybrid soil conditions are relatively rare, the hybrid soils are less abundant than the parents. These morphological oak species are neither genetically or reproductively isolated. But their hybrids are unadapted to the existing edaphic conditions and cannot become established.

Morphologically distinct species that grow within pollination range of each other and that are not reproductively isolated but are ecologically specialized in relatively common among related species of long lived woody plants. Other familiar examples are *Salvia mellifera, Salvia apiana* (Epling 1947, Grant and Grant 1964) certain species of *Betula* and *Populus* (Barners 1961) and *Arctostaphylus* (Gottlieb, 1968).

Table 9.0

Synoptical List of Small-Flowered Cobwebby Gilias Grouped According to Their Morpho-logical and Cytogenetic Relationships into More or Less Natural Species Groups.

[From Grant, 1964, by Permission of Author and Publisher.] See Also Fig. 9.1

A. Gilia ochroleuca Group (Diploids)

1. *Gilia ochroleuca ochroleuca* Jones. Mojave Desert, California
2. *Gilia "obispo."* South Coast Ranges and San Diego Co., California

3. *Gilia mexicana* A. Grant and V. Grant. Desert grassland; Arizona, New Mexico, and northern Chihuahua, with a disjunct area in northern Baja California
4. *Gilia clokeyi* Mason. *E.* California to NW New Mexico
5. *Gilia "bradbury."* Death Valley, California
6. *Gilia aliquanta* A. Grant and V. Grant. Mojave Desert, California, and SW Nevada

B. Gilia tenuiflora Group (Diploids)

7. *Gilia interior* (Mason and A. Grant) A. Grant. California and Nevada
8. *Gilia austrooccidentalis* A. Grant and V. Grant. Inner South Coast Ranges, California
9. *Gilia jacens* A. Grant and V. Grant. Inner South Coast Ranges, California
10. *Gilia minor* A. Grant and V. Grant. Central California to W. Arizona
11. *Gilia salticola* Eastwood. Eastern Sierra Nevada and western Great Basin, California and Nevada.

C. Gilia brecciarum Group (Diploids)

12. *Gilia brecciarum brecciarum* Jones. Oregon and Nevada to southern California
13. *Gilia diegensis* (Munz) A. Grant and V. Grant. Southern California

D. Gilia inconspicua Group proper (Tetraploids)

14. *Gilia tweedy!* Rydberg. Oregon to Wyoming
15. *Gilia inconspicua* (Smith) Sweet. Washington to northern Nevada and Utah
16. *Gilia transmontana* (Mason and A. Grant) A. Grant and V. Grant. Mojave Desert, California to western Utah
17. *Gilia malior* A. Day and V. Grant. California and Nevada
18. *Gilia flavocinta austrail's* (A. Grant and V. Grant) A. Day and V. Grant. Arizona and New Mexico

19. *Gilia ophtalmoides* Brand. Eastern California to Colorado and New Mexico

E. Gilia sinuata Group (Tetraploids)

20. *Gilia sinuata I (G. sinuata* Douglas). Washington to western Colorado and southern California
21. *Gilia sinuata II.* Arizona
22. *Gilia modocensis* Eastwood. Eastern Oregon through Nevada and southern California
23. *Gilia tetrabreccia* A. Grant and V. Grant. South-central California Ft **Gilia crassifolia Group** (Tetraploids and Octoploids)
24. *Gilia crassifolia 4* x . Argentina and probably Chile
25. *Gilia crassifolia 8* x . Southern Argentina (Patagonia)

Correlation between Genetic Isolation with Morphological Identical Species

Genetically isolated populations that are morphologically identical or very similar. Grant(1964) thoroughly investigated the Gilia inconspicua complex of species. The complex ranges through out the arid regions of Western North America and Southern South America where the plants occur in open sandy places in a variety of habits. They can be found growing in desert washes or sandy river flood plains; in openings of the desert scrubs vegetation or on the slopes of mountains.

The individuals are self compatible and the pollen is deposited on the stigma lobes automatically. The products of self pollination are fully vigorous generation after generation as was observed in experimental progenies. Grant (1964).

At maturity the plants belonging to the *Gilia inconspicua* complex bear small, dull coloured flowers on the upper branches. In their general characteristics although they may look alike, they also differ in various fine details of external morphology. Based on detailed morphological, cytological and crossing experiments it was possible to demonstrate that it was not one species, but was actually a group of six independently derived groups of species comprising no less than twenty five species as shown in table 9.0.

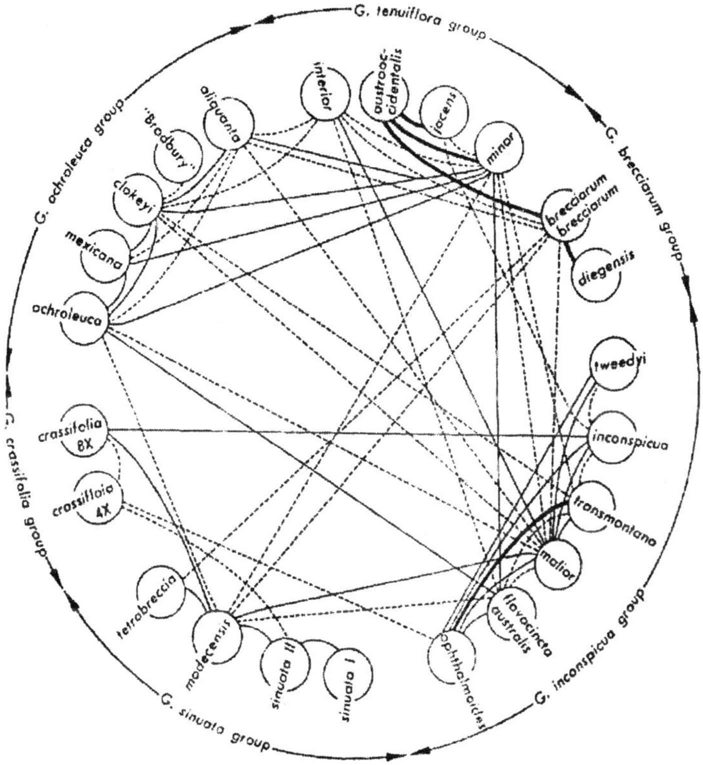

Fig. 9.2. Crossability and sterility barriers between twenty-three sibling species in the *Gilla inconspicua* species complex. See Table 5-1 for geographical location, chromosome number, and taxonomic arrangement of the species. (From Grant, 1964, by permission. Copyright Academic Press, 1964.)

With a few exceptions, all twenty five species of these species are completely genetically isolated from each other, no hybrid seed having been formed inspite of repeated attempts at crossing as shown in figure 9.2. In a few cases where hybrids were produced, they were very sterile. (highest fertility 28% of fertile of pollen. The patterns found in these species of *Gilia* is known from other groups of annuals known as *Clarkia* (Lewis 1962). *Oenothera* (Raven). Reproductively isolated species that are morphologically indistinguishable or nearly so are called sibling species. The most important lesson that we should learn is that there is no absolute correlation between morphological differences and reproductive isolation. Although in most cases generally different morphologically differences are correlated with reproductive isolation, this is not in all cases. Speciation is the process by which new species are formed from other ancestral ones through modification in different geographical regions.

Modes of Speciation

There are two modes speciation by which new species can originate

1. Phyletic (Transformation): One species is transformed into another by transformation slowly and steadily In course of time one species acquires some adaptive characters and transforms into another.
2. Geographical theory of speciation or **Allopatric** theory of speciation: In this one ancestral species gives rise to one or more species in different geographical regions without necessarily loosing its identity. This process is referred to as multiplication of species or true speciation, allopatric speciation.

The Process of Dynamics

Multiplication of species or true speciation is what has been called the geographical theory of speciation or Allopatric theory of speciation. According to Moritz Wagner (1868) the 1st step of speciation is reproductive isolation brought about by physical (geographical) separation. Second step is the independent evolution of these reproductively isolated populations. If and when the ranges of the populations eventually merges again mechanism such as genetic incompatibility has evolved, the speciation process is considered completed.

Dynamics of the process: Four steps can be identified:

In the separation of the original gene pool into two or more isolated gene pools as a result of changes in the environment. The separation of genepool results commonly from a shrinkage of the species so that population become physically completely reproductively isolated.

This separation may also be the result of long range dispersal when new populations become established beyond the range of pollination of the original gene pool. Both these phenomenon are common and almost every species with an extended range has populations that are reproductively isolated from the rest of the population of the species.

The process of geographical isolation affects all kinds of plants. Species that are pollinating have a greater chance of successful establishment after long range of dispersal. In this way plants are adapted to disturbed or open habitats.

Differentiation

The process by which two populations in isolation evolve in different directions, so that they become morphologically and physiologically different. Differentiation does not necessarily follow upon isolation, although apparently it often does. There seems to be ample of evidence for the idea that population will adapt closely to the conditions of the environment in which they are found. Ecotype information is probably the 1st step in differentiation as shown in figure 9.3.

Bradshaw (1959), found a grass which is self incompatible-perennial grass that grown in a wide variety of habitats in temperate regions. *Agrostis tenuis* are also found growing on mine workings in the vicinity of abandoned mines. These mine workings comprises high lead concentrations and toxic to most vegetation. Very few plants used to grow on them including plants *Agrostis tenuis* form normal soil. Populations have evolved by lead tolerance (Bradshah, 1960). Tolerance does not appear to be included by conditioning processes, is not lost in cultivated in the absence of metal, and is heritable (Bradshah, 1959,Bradshah, Mc Neilly and Gregory, 1965).

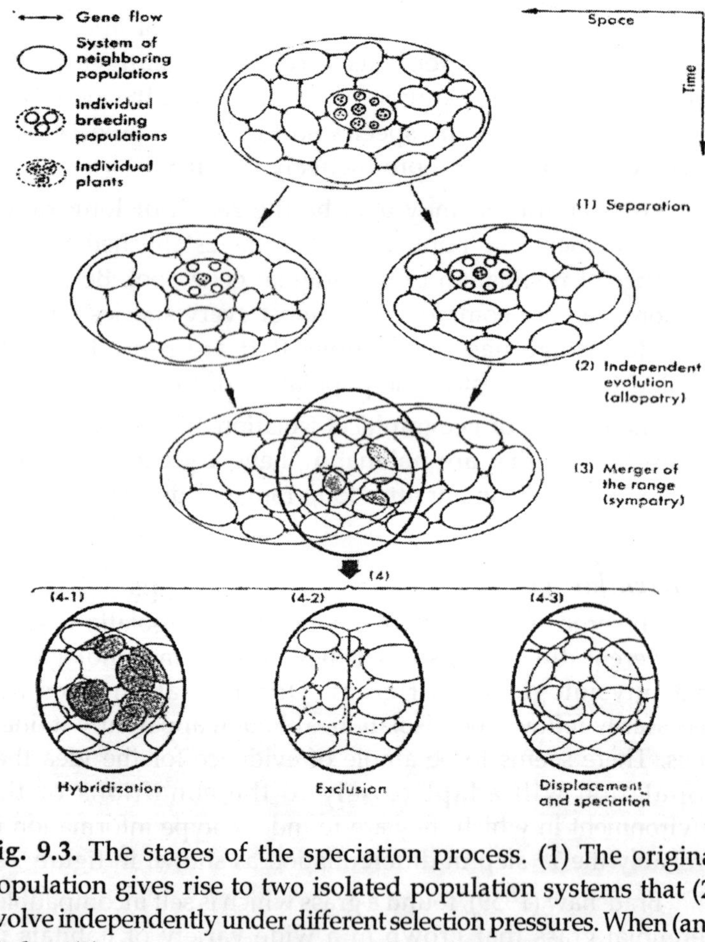

Fig. 9.3. The stages of the speciation process. (1) The original population gives rise to two isolated population systems that (2) evolve independently under different selection pressures. When (and if) they (3) merge their range partially or entirely, they may (4-1) hybridize, restoring the original situation of one species; or (4-2) they may exclude each other, restoring the case of two geographically isolated population systems; or (4-3) they may displace and form the new situation of two sympatric species. This last case can also occur if the species have differentiated before they merged their range.

In the population study of Goginon mines in wales, Bradshah could demonstrate that the transition zone between lead tolerant and intolerant populations is less than 29 meters wide. On each side of the zone have the same levels of tolerance and intolerance as the plants considerably further away from the boundary. Analysis of the ecological and historical factors indicate that the lead tolerant forms must have developed recently and that they have evolved in a place that is within a few feet from the nontolerant forms and within population range. However the selection pressure is so strong that the hybrids have no where to become established. They are not able to tolerate the high lead content. On one side of the line or the competition from the other plants on the other. Lack of hybrid survival contribute in this case an absolute or almost absolute barrier to gene exchange. Similar situations with varying degree of intermediacy have been reported for species adapted to serpentine and other soils. *Strephanthus* and other species (Kruckeberg 1951,1954,1957,1958) *Erigeron* (Mooney, 1967) and *Arctostaphylos* (Gottleib, 1968). In all these instances there are two populations a serpentine tolerant and intolerant one. No absolute isolating barrier exists between them. The hybrid establishment is not possible as it is restricted to a narrow zone. In these plants close adaptation of plants to the environment was pointed out and introduce new factor, the role of disruptive selection. In case of *Streptanthus* and *Arcostaphylos* the serpentine and non serpentine plants differ morphologically and it cannot be ascertained whether the present contact has always existed or is of recent origin due to expansion of the range of either or both species. But in *Agrostis* and probably also *Erigeron* the evidence points to differentiation having taken place without geographical separation. This is very important information since this point has not been and is still being discussed vigorously.

Merger of the Range

The forces that lead to an overlapping of the ranges of two hither to isolated populations are similar to those that lead to their separation. Basically these forces are environmentally changes that lead to the expansion of the range or migration of the population or population systems into an area previously not occupied. It is also possible that evolution of the populations produces physiological or morphological changes (a better seed dispersal apparatus) that are directly or indirectly responsible for expansion as shown in figure 9.3. But this step is not a necessary condition in process of speciation.

Range Overlap and Resulting Competition

The superposition of the range of the populations so that their members are within population range and the resulting interactions is the last step in the speciation process. The two populations that come together may have differentiated physiologically and genetically so that they will have different ecological requirements and that their gametes will no longer fuse to form viable zygotes. In such case speciation process has been completed without interactions as shown in fig 9.4(a).

If the populations that come into contact have the same ecological requirements and are capable of hybridizing they will compete with each other. The result of this competition can have one of the three out comes.

1) One species out competes the other and the previous non overlap of the range is restored.
2) As a result of selection of the two species divide the environment in such a way that they no longer compete and at the same time some kind of breeding barrier is established so that few or no hybrids are formed.
3) The two populations hybridizes and form on interfertile population.

Exclusion Principle

Whenever two populations share the requirement for some or all of the environmental components in a way that causes mutual interference, the two populations are competing for these resources. When two populations are competing they tend to displace each other, i.e., the exact requirements of one or both populations will change in such a way that the competition is lessened as shown in fig 9.4(b).

Table 9.1

Genotypic Composition in Per Cent of the Total of Two Populations After One Generation of Reproduction Assuming No Fitness Difference Between Genotypes [see Text for Further Explanation]. In One Case All the Hybrid Seed Germinates and Forms Fertile Plants; in the Other Case No Hybrid Seed Germinates.

Genotype	Initial Composition		No Postfertilization Barrier			No Hybrid Seed Formed	
	Pop. 1	Pop. 2	Pop. 1	Pop. 2	Hybrids	Pop. 1	Pop. 2
aabbcc	0.2401	—	0.16807		0.0046	0.6561	—
aabbCc aaBbcc Aabbcc	0.4116	—	0.2881		0.0279	0.2916	—
aabbCC aaBbCc AabbCc aaBBcc AaBbcc AAbbcc	0.2646	0.0081	0.18522	0.0056	0.0563	0.1944	0.0001
aaBbCC AabbCC aaBBCc AAbbCc AaBBcc AABbcc	0.0756	0.0756	0.0529	0.0529	0.1125	0.0360	0.0360
aaBBCC AaBbCC AaBBCc AABbCc AAbbCC AABBcc	0.0081	0.2646	0.0056	0.18522	0.0563	0.0001	0.1944
aABBCC AABbCC AABBCc	—	0.4116	—	0.2881	0.0279	—	0.2916
AABBCC	—	0.2401	—	0.16807	0.0046	—	0.6561

seeds are absolutely sterile and consequently do not germinate, the effective populations, that is, the ones that are going to reproduce the two populations, are formed largely by the plants outside the area of overlap. In these plants the gene frequencies are as follows: population 1 (tolerance range 30–50): $A = 0.0$; $a = 1.0$; $B = 0.1$; $b = 0.9$; $C = 0.1$; $c = 0.9$. Population 2 (tolerance range 80–100): $A = 1.0$; $a = 0.0$; $B = 0.9$; $b = 0.1$; $C = 0.9$; $c = 0.1$. The change in the gene frequencies has been quite drastic in the direction of a fixation of the recessive genes in population 1 and the dominant in population

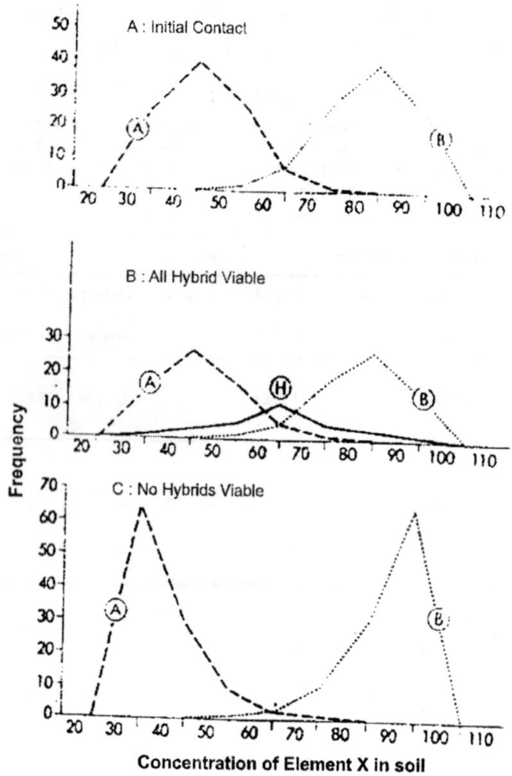

Fig. 9.4. Distribution of two hypothetical populations A and B with relation to an element X in the soil. Both populations are interfertile; they have, however, differentiated in isolation with relation to their tolerance to the element X. The assumptions are that the frequency of hybridizations is proportional to the degree of overlap of the populations, that breeding is random in the zone of overlap and that all pollinizations produce seed. (A) Initial frequency of the two populations in relation to the concentration of the element X. (B) Frequency of the two polulations after one generation assuming all hybrid seeds to be fertile. No displacement has occurred; the proportion of parental types in relation to the previous generations has decreased however. (C) Frequency of the two populations after one generation assuming all hybrid seeds to be inviable. Maximum displacement away from the zone of overlap has taken place.

Displacement

Two semi differentiated populations coming into contact may hybridize, and the hybrids, if not totally sterile, will in turn back cross with the parental populations. Two populations have partially differentiated in relation to their tolerance to certain harmful environmental factors as shown in fig 9.4(c).

Conclusion

Speciation is the development of populations of freely interbreeding organisms adapted to their environment that are reproductively isolated from other such populations. Present day species are evolutionary products that have descended from preexisting species. Variation in organisms provides the raw genetic material that in concert with environmental variation makes speciation possible. Variation in natural populations is limited by such factors as self fertilization, as external reproduction, genetic fixation and stabilizing selection. In general, many more individuals are produced by each generation than will survive and reproduce. Variation does occur in each generation and some individuals within the population will have better survival characteristic that are heritable. The biological species represents the major continuity in on going evolution. In all geographically isolated populations or groups of populations, which have only or no gene exchange with other populations of the species. Any insular population is normally such as isolate and isolates are particularly common near the periphery of the species of range. Isolates are frequently of sufficient differences to merit subspecies rank. The biological importance of the geographical isolate in the every isolate regardless of its taxonomic rank, is an incipient species; it is an important unit of evolution.

Types of Species

Types of species	Units being classified	Sphere in which applicable
1. Taxonomic	Taxa: groups of morphologically Similar individuals	Groups with morphological differences that are useful in formal classification.
2. Biological	Sexually reproducing population systems.	Sexual organisms on a single time level.
3. Micro	Populations in uniparental organism	Uniparental organisms including asexual and Parasexual organisms.
4. Successional	Phyletic lineages	Biological species and microspecies
5. Biosystematic	Fertility groups	Sexual plant groups that have been artificially hybridized in the experimental garden.
6. Evolutionary	Combined sexually reproducing Populations, uniparental groups, and phyletic lineages.	Composite entity of natural units.

Taxonomic Species

The taxonomic species, from a practical standpoint, is an assemblage of morphologically similar individuals that differs from other such groups in one or more structural characteristics. The taxonomic species is the usable unit in phenetic classification and identification.

Biological species

Biological species is a group of interbreeding populations

composed of biparental organisms reproductively isolated from other breeding populations. Grant (1981) states that this basic proportional unit can be viewed as a stage in a more or less continuous series of levels of divergence in population systems.

Semispecies

Semispecies are population systems in an intermediate stage of gradual and continuous Evolutionary divergence from a geographic race to a sympatric species. Disjunct allopatric populations of *Liriodendron chinense* and *Liriodendron tulipifera* that live in similar habitats in China and the southeastern United States are interfertile when experimentally crossed.

Microspecies

Microspecies (Grant, 1981) fall into four main classes according to the mode of reproduction: 1) clonal microspecies, reproducing by vegetative propagation, as in Phragmites;2) agamospermous microspecies, reproducing by agamospermy, as in *Rubus* spp.; 3) heterogamic microspecies, reproducing by the *Oenothera biennis* or *Rosa canina* genetic system; and 4) autogamous microspecies, predominantly autogamous and chromosomally homozygous, as *in Erophila.*

Successional species

Successional species (Grant, 1981) are members of the same lineage living in different geologic time periods. They are phyletic species with different sets of characteristics. Phyletic evolution is a continuing process with intergrading forms connecting distinctive species at different points in evolutionary time.

Biosystematic species

Biosystematic species (Grant, 1981) are classified as categories based on fertility relationships as determined by artificial hybridization experiments. These biosystematic categories form a hierarchy of ascending units; ecotype, ecospecies, coenospeices, comparium.

Evolutionary species

Evolutionary species as defined by Simpson (1961) in Grant (1981) is a population that possesses the following characteristics.

1. It is a lineage, an ancestral-descendant sequence of populations existing in space and time.
2. The lineage evolves separately from other such lineages, that is, from other species.
3. It has its own 'unitary evolutionary role,' which is to say that it fits into its own particular ecological niche in a biotic community.
4. It has its evolutionary tendencies, being susceptible to change in evolutionary role during the course of its history.

10

Polyploids, Abrupt Speciation and Catastrophic Selection

Objectives

1. To understand the doubling of chromosomes consequent of inhibition of gene exchange leads to genetic isolation.
2. To assess how allopolyploids are having advantage over the autopolyploids in speciation
3. To understand in what arid conditions the selection favors the genetic alteration of chromosomes
4. To know how interchanges and inversions occurred in those *clarkia* species have altered genetic make up useful for genetic isolation selection.

Abrupt Speciation Definition

It is the sudden change in chromosome number constitution producing instantly an almost irreversible barrier between populations and thus effectively isolating them. The abrupt change is generally the result of polyploidy. Speciation of this sort is initially sympatric in that the new eco species and its progenitor (s) have coincidental or least over lapping distribution. Polyploids may however, if successful, come to occupy distinct eco-geographical areas why the general incidence of polyploidy increases with latitude and altitude, and to be certainly the cause of frequency of polyploidy being much higher in old groups than in young ones.

Among plant species at times speciation occurs abruptly rather than by the gradual process. This occurs as a result of the doubling of the chromosomes number of the plant, a process called **polyploidy**.

The Species Problem and Classification

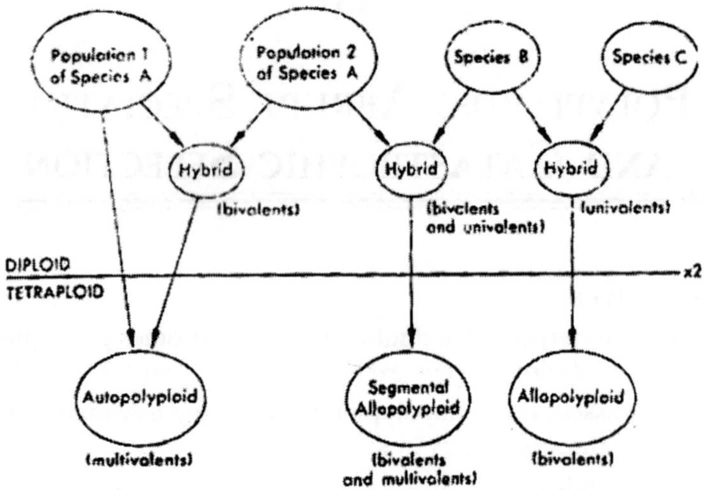

Fig. 10.1. The various types of polyplidy. Autopolyploidy results from the doubling of a plant, or an *intraspecific* hybrid. Because autopolyploids have four sets of identical or nearly identical chromosomes, they are expected to form associations of four (quadrivalents) at meiosis, and as a result they are expected to be sterile. Allopolyploids result from the doubling of the chromosome s of an *interspecific* hybrid between two relatively unrelated species. The diploid hybrid is sterile because the chromosomes of the two speciesare nonhomologous and will not pair; however when the chromosome s are doubled, bivalents can again be formed and the allopolypoid is halfway between the autopolyploid and an allopolyploid, in that the chromosomes of the two species are related enough to form bivalents in the diploid hybrid, but also form bivalents at the tetraploid level.

Polyploidy can be defined as the phenomenon that results in doubling of the number of chromosomes of a plant. The resultant plant in most cases is incapable of forming fertile offspring with members of the ancestral diploid population. It has aquired instant reproductive isolation and if it is capable of reproducing and of finding an ecological situation that it can exploit, a new species has been formed as shown in fig 10.1. Among the Angiosperms close to 40% of all species appear to be polyploidy, so that polyploidy is a very important phenomenon.

Different Polyploids

Polyploids are basically of two kinds 1) autopolyploids which are the result of doubling of the chromosomes of a normal plant. 2) Allopolyploids, which are the result of doubling of the chromosomes of hybrid plant. This latter kind is apparently more frequent.

Formation of Allopolyploidy Species

The probability of establishment of an allopolyploid is higher for cytological reasons (meiosis is more likely to be normal) and also because the hybrid may have novel characteristics of the two parental species. (meiosis is more likely to be normal) and also because of the hybrid may have novel characteristics that will allow it to exploit habitats not open to either of the parents. In autopolyploids the interchanges or association of four or six or eight will be formed. When these associations are present in those plants, they exhibit the chromosomal sterility. Hence autopolyploids they form only embroidery to those plants. It is only allopolyploids, in this homologous chromosomes form perfect bivalents such as ring, rod and 'I' shaped bivalents. Such plants can exploit more vigorously the habitat and tend to develop new heritable adaptive characters with which such plants can survive very well.

From the taxonomic point of view polyploidy presents species problems. Autopolyploids are morphologically very similar to the parental population. Allopolyploids are intermediate or combine the characteristics of the two parental species. In the latter case blurring of differences between species occurs and in the former case a non correlation between morphology and isolation exists.

Catastrophic selection

In some instances very rapid speciation can take place leading to genetic isolation with little no morphological differentiation but *without polyploidy*, a phenomenon known as Catastrophic selection (H. Lewis, 1962).

Catastrophic selection occurs in ecologically marginal populations. In *Clarkia* (Onagraceae) has observed how exceptional drought by eliminating all but one or a few deviant individuals, has lead to rapid speciation at the diploid level. This is an extreme example of aridity as evolutionary stimulus (Stebbins, 1952)

Biosystematic Studies of H. Lewis

The best examples provided by the genus *Clarkia* of the family Onagraceae. *Clarkia* is a group of plants found in the western United States centered around California. Harlan Lewis and his students made intensive studies of biosystematics of the genus *Clarkia* for twenty years. In that study interspecific crosses have been attempted between most of the species. The cytological analysis of the hybrids so obtained showed that many species differ quite drastically in the linear order of the genes.

Species From Californean Range

The cytological analysis of the hybrids so obtained showed that many species differ quite drastically in the linear order of genes. In the process, interspecific crosses have been attempted between most of the species. Lewis and Raven, 1950 explained the situation by *Clarkia franciscana*. This species known to occur only in one locality a serpentine covered slope overlooked in the Golden Gate at the East end of the Presidio in San Fransisco, California. The species is morphologically similar with two other species *C. rubricanda* and *C. amoena*, that occur close by as shown in fig 10.2.

The first of these species is found along the coast and in the valley of the California coast ranges from northern San Luis Obispo County to just north of San Fransisco Bay in Marin county. *Clarkia amoena* replaces *C. rubricanda* in central Marin county and continuous northward along the coast to British Columbia and Vancouver Island.

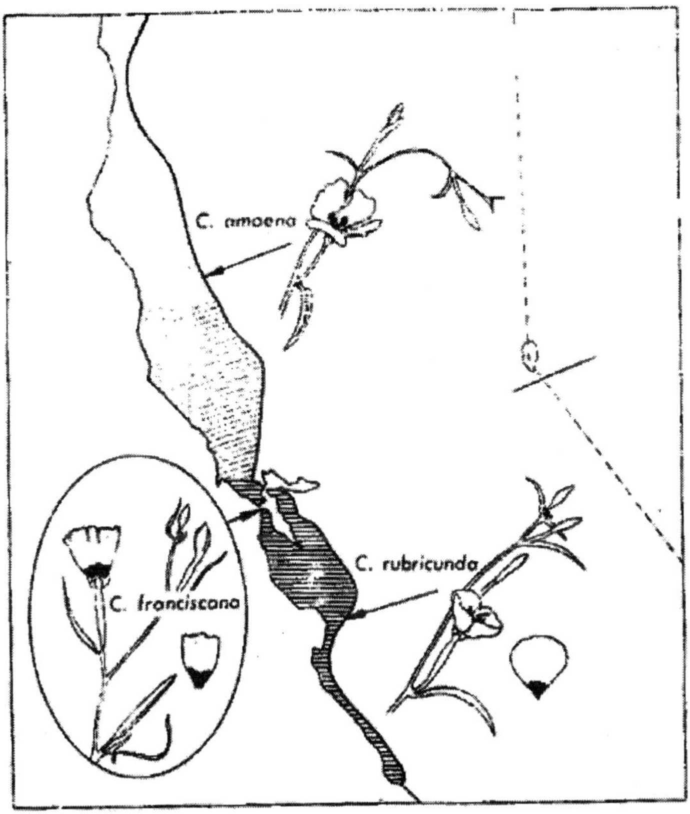

Fig. 10.2. Distribution of *Clarkia amoena. C. franciscana,* and *C. rubricanda* in California. Note how morphologically similar the three species are. The best distinguishing feature is the petal, particularly the spot, which is bright red and present near the center in *C. amoena,* is well defined, conspicuous, reddish purple and basal in *C. franciscana,* and is not so well defined and merges somewhat in *C. rubricanda.* (From H. Lewis, 1958, by permission.)

Chromosomal arrangement among the Hybrids of *Clarkia* Species

Raven and Lewis attempted to produce interspecific hybrids among Clarkia francisiana, *C. amoena* and *C. rubricanda,* in several combinations to determine the fertility of hybrids and to study the differences in chromosome architecture indicated by chromosome pairing at meisosis. The fertility of the pollen was not more than 2%. The pairing of the chromosomes in the hybrids showed that the chromosomes of these species differed drastically in the arrangement of homologous segments. *Clarkia fransciscana* differs from *C. rubricanda* by at least two translocations and two or more paracentric inversions as well as by a series of smaller rearrangements. *Clarkia amoena* in turn, differs from *C. rubricanda* by three translocations and two inversions. The chromosomal changes are of an entirely different magnitude than in the case of most species such as *Glandularia.*

Generation of Genetic Isolation

When two closely related species or populations differs in the order of the genes in their chromosomes, it is obvious that in one or both of the differing species or populations changes have taken place in the architecture of the chromosomes because of the common ancestor of these species or population must have had one arrangement only (at least at one time) How does this takes place? The usual assumption is that a small chromosomal changes takes place in a plant of population that gives that plant in higher fitness. In time all plants of the population will have the new chromosomal arrangement. By the repetition of this process the populations can become quite distinct cytologically. These changes presumably are accompanied by morphological and physiological changes as the populations become adapted to different ecological conditions.

Formation of Heterozygous Genetic Architecture

How the great dissimilarity in the chromosomal

arrangement of the genes between *Clarkia franciscana, C. amoena, and C. rubricanda* and their great morphological similarity suggest that they have originated in ways other then by the gradual accumulation of small changes. Whenever the arrangement of the order of the genes in a chromosome of a plant is changed, crossing over within the altered segment in a plant that is heterozygous for this change will produce two sterile gametes out of the four gametes that are the normal product of meiosis. If the altered segment is small, the probability of a cross over occurring within the segment is very small, consequently a cross over will not occur in most cells undergoing meiosis, and the new arrangement can become established in the population. However if the segment is large, the converse is true. Therefore the probability that the chromosomal arrangement of *Clarkia franciscana* are the result of small accumulation of small changes is indeed very low because the repeated establishment of small chromosomal changes without corresponding changes in morphology is an improbable event.

Hypothesis About Genetical Isolation

H. Lewis 1962; Lewis and Raven, 1958 therefore hypothesized a mechanism that he termed catastrophic selection. According to this hypothesis, under certain unknown conditions (environmental stress, drought) or mutator genotypes (genotype) that produces a high frequency of mutations have been mentioned as possible causes. Drastic chromosomal rearrangements take place in a plant. If a plant with such a drastic chromosomal rearrangement becomes detached from the population it can establish a population with a new chromosomal arrangement that is genetically isolated from the ancestral population from its very inception.

What forces bring about such drastic changes in some case and so little in others is still a mystery. Particularly because some species that have been isolated for millions of years have not developed any genetic or chromosomal differences.

Hybrids Produced From Two Separated Species For Long Period

A very rare case suits the best example for the one of the aspect of the above. *Platanus orientalis, Platanus occidentalis.* The first of these species is a native of the Middle East Asia and the second of Eastern North America. They are morphologically very distinct and also physiologically so, because they grow in different climatic regions. For how long these two species have been isolated is impossible to say with any certainty but it is likely a million or more years. The artificial hybrids between these species known as *P. acerifolia* and widely cultivated as an ornamental tree is vigorous and highly fertile and has a normal meiosis indicating that no drastic changes in the order of the genes has taken place.

Conclusion

The only general conclusion that can be drawn is that when two populations are isolated, physiological, morphological and genetic differentiation proceeds differently in each cases. The general terms 1) the degree of environmental changes to which the species adjusts and 2) genetic system of each species, which controls and determines the way a species responds to the environmental changes.

Biological Species Concept

Groups differing in chromosome number between which there are well defined ecological and geographical differences and which are capable of only limited gene exchange.

Biological species concept: Speciation is the development of populations of freely interbreeding organisms adapted to their environment that are reproductively isolated from each other such populations. Present day species are evolutionary products that have descended from pre existing species. Variations in organisms provides the raw genetic material that in concert with environmental variation makes speciation possible. *'The biological species'* is a group of interbreeding population composed of biparental organisms, reproductively isolated from other breeding populations. According to Grant (1981) this basic unit can be viewed as a stage in a more or less continuous series are local races, geographical races, allopatric species.

Local races are local breeding populations that differ from other conspecific populations in composition of their gene pools and in their genetically determined phenotypic characters. Geographic races represent local racial variation over a wide range. The geographical variation may be continuous and the races are disjunct if the habitat is discontinuous, such as alpine races on different mountain peaks. Geographical races are allopatric in that they occupy different geographical areas.

The geographic races may also be ecologic. The climatic races of *Potentilla glandulosa* complex from the alpine zone of sierras of California, the same mountain bloom earlier in a uniform garden that mid altitude races from the same mountain (Clausen, Keck, and Hiesey, 1940, 1948). The edaphic races of *Grandularia capitata* and *Strephanthius glandulosa* have been shown by Kruckberg (1951) to be both serpentine tolerant and serpentine intolerant.

11

Leaky Isolation Barrier, Gene Flow and Selection

Objectives

1. To understand the two important isolating mechanisms having significance
2. To assess the establishment of Leaky isolation mechanism through allochronic selection.
3. To understand the restriction of gene flow
4. To understand and assess the free gene exchange

Definition

Two very important isolating barriers are 1) the gametic incompatibility, that is the ability of the pollen to germinate on the style of the sperm of one species to form a viable seedling with the egg of the second species and 2) an absolute hybrid sterility, the incapacity of the hybrid to form gametes. All other barriers are likely to break down and may be termed "leaky isolating barriers". Species that are geographically separated may come together in the course of time; some pollen grains may in some instances germinate on the style of the supposedly incompatible species inbreeders and apomicts occasionally outbreed and hybridize; the tail end of the blooming period of one species may overlap with the beginning of the blooming period of second species; and hybrids germinate and grow in areas where the environment is intermediate between that of the two parental species, even in cases where the hybrids are not vigorous.

Table 11.0

ISOLATION MECHANISM

Prefertilization	Reduction of contact	1. Geographical separation 2. Ecological separation
	Reduction of mating frequency	3. Pollen incompatibility* (gametic isolation) 4. Inbreeding and asexual reproduction 5. Specific pollinators (ethological isolation) 6. Different flowering times (allochronic isolation)
Postfertilization	Reduction of zygote formation (prezygotic)	7. Gametic incompatibility*
	Reduction of gene flow through hybrids (postzygotic)	8. Hybrid sterility* 9. Hybrid weakness or breakdown* 10. Lack of hybrid establishment (environmental isolation)

* May be absolute and irreversible.

Two Important Isolation Mechanisms

Major isolation mechanisms found among plants 12 out of these the only truly absolute isolating barriers are 1) gametic incompatability, that is the inability of the pollen to germinate on the style or of the sperm of one species to form a viable seedling with the egg of the second species and 2) absolute hybrid sterility, the incapacity of the hybrid to from vaiable gametes. As we have seen once hybrids are established they are capable of crossing with the parental species, leading to the formation of back crosses and gene exchange between populations.

Leaky Isolation Mechanism Establishment

Because so many plant species are separated by leaky isolating barriers and consequently may hybridize, this is there not fallacy in our species concept. In effect, a strict application of the biological species concept to plants will lead to circumscription of only those taxa that possess absolute barriers to hybridization. This means that many species of oaks would have to be considered as one species as well as practically all birches and willows, many ferns, orchids and grasses and so on, obviously they are not the same but different species. Consequently, even in cases where no absolute isolating mechanisms are present new species can be formed and maintained.

Gene Flow and Selection

Introduction

The basic unit in the evolution is gene pool which is equated with the local breeding population. Even the concept of gene pool is an abstraction to certain extent because it cannot be defined in strict scientific terms. The breeding mechanism has a great effect on the size and variability of gene pool in some out breeding species which have effective pollinating agents. In such out breeding population the gene pool may be large at the other extreme, but in some population of apomictic plants may hardly exchange genes at all generation after generation.

Gene Flow Restriction

Any mechanism that restricts gene flow may favour speciation. The converse viz when two populations are subjected to different selective forces. They will tend to diverge genetically, physiologically and morphologically. Gene exchange will slow down the process of divergence because intermediate forms will be constantly formed. If selection is not sufficiently great, the formation of two sharply distinct population is not possible. Nevertheless if selection is intense and the intermediate zone is very narrow, the formation of two distinct populations can occur and has occurred. Good example are the evolution of population adapted to extreme edaphic conditions, such as serpentine.

Gene Exchange Barrier

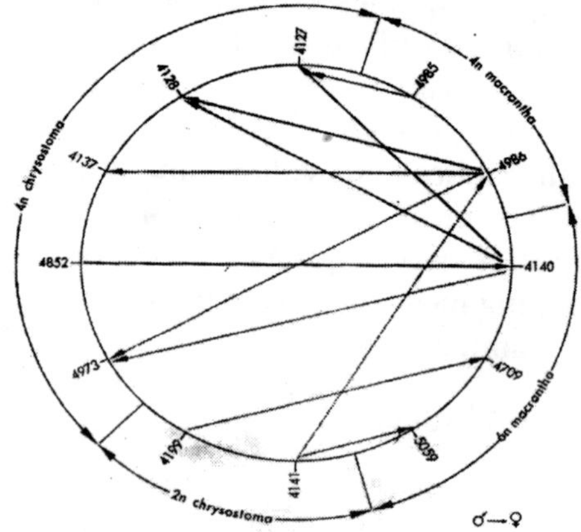

Fig. 11.1. Diagram summarizing average pollen viabilities of progenies obtained from artificial interspecific hybridizations in *Lasthenia*, section *Baeria*. (From, Ornduff, 1966, by permission.)

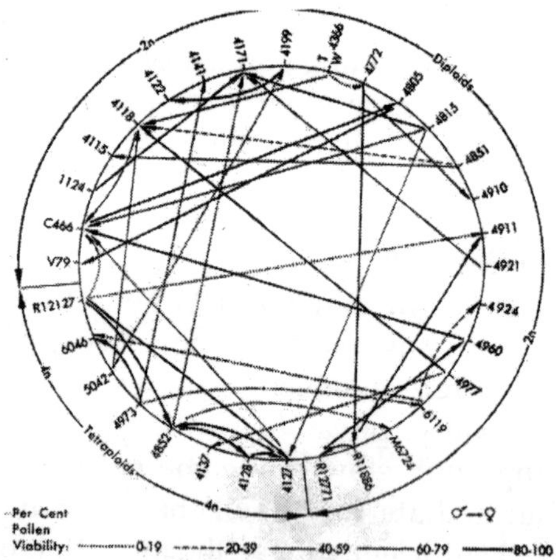

Fig. 11.2. Diagram summarizing average pollen viabilities of progenies obtained from intraspecific hybridizations within and between diploid and tetraploid races of *Lasthenia chrysostoma*. (From Ornduff, 1966, by permission.)

The absolute barrier to gene exchange exist with little or no morphological differentiation. The famous example for this is *Gilia inconspicua* species group studied by Grant (1964). Another example is *Lasthenia* belonging to the family compositae. According to Ornduff (1966) found in a section Baeria for example, *Lasthenia chrysosthoma* was found to have diploid and tetraploid races as shown in figure 11.2. While the other species *L. macrantha* has diploid and hexaploid races. Crosses between populations of *L. chysothoma* with different chromosome numbers yielded hybrids that has reduced pollen fertility, although occasionally the fertility was as high as 50%. However, hybrids between diploid populations also often had reduced fertility, in one instance less than 20%.

Free Gene Exchange Barrier

Crosses within *L. macrantha* on the other hand to be highly fertile even where the chromosome number differed. Finally, interspecific hybrids were relatively fertile or were sterile according to the populations crossed and the chromosome number of these populations, but they usually had better than 50% fertility as shown in figure 11.1.

Space and Time

In nature there are some plant species that are not separated by absolute isolating barrier and equally important that there are species whose populations are not entirely interfertile. In nature most of the populations of a species are isolated from other populations because of space and time.

Seed and Pollen Transportation

Gene exchange between a populations of *Glandularia canadensis* from Nashville, Tennesse and one from Gainsville, Florida must be very small indeed. The pollen from one population to the other distantly placed populations, seed exchange between the populations is so small as to be almost nil, any exchange is through geographically intermediate populations. Such a process would take at best tens or hundreds of generations. Experiments with these plants show, the Nashville, Tennessee, and the Gainsville, Florida, populations have not acquired reproductive isolation and although they have differentiated morphologically, the differences are minor.

12

Isolating Mechanisms

Objectives

1. To understand the isolation mechanisms, how the mechanisms of isolation develops.
2. To understand prezygote mechanisms and post zygote mechanisms which operate before or after sexual fusion.
3. To understand which two of the isolating mechanisms are important.

Introduction

The adaptively valuable gene combinations of each species can be broken up by hybridization. Hybrid breakdown is a manifestation of this disintegration. The other reproductive isolating mechanisms serve to protect the adaptive gene combinations from disintegration. Properties of individuals that prevent successful interbreeding with individuals that belong to different populations. The reproductive isolation of biological species, the protection of its collective gene pool against pollution by genes from other species, results in discontinuity not only of the genotype of the species but also of its morphology and other aspects of the phenotype produced by this genotype.

Since the evolution and operation of various isolating

mechanisms obviously affect the degree of phenotypic distinction between population from which taxonomists construct their taxa, it is important for us to consider examples of these mechanisms and their bearing on the species and infraspecific units.

The various factors causing isolation between populations normally build up gradually and therefore we consider briefly the distinction that is made between ecospecies and races. The genecological differentiation of populations giving rise in many instances to ecotypes and the geographical differentiation of populations giving rise to subspecies. These are infra specific products and are often referred to as races, ecological or geographical. They may be distinguished taxonomically as subspecies or varieties or the differentiation may take on a continuous (clinal) pattern.

Isolation caused is a key factor in speciation in allowing new evolutionary lines to be built up and maintained. Detailed discussion of isolating mechanisms, largely for an evolutionary point of view have been published by Clausen (1951), Stebbins (1950).

How Isolating Mechanisms will occur

Isolating mechanisms

The facts that many hybrids which does not occur in the wild have been synthesized experimentally and that hybrids known in the wild do not always occur wherever their parents grow close together; show that there must be factors operating in the wild which prevent hybridization. It is possible to divide these according to whether they operate before or after sexual fusion mechanisms and post pollination mechanisms. Alternatively mechanism 1-7 may be described as external and 8-15 as internal. The following classification follows that of Stace 1975.

Geographical Isolation

Two species are separated geographically so that they do not come into contact. Examples: *Laryx deciduas* in Europe and *L. kaempferi* in Japan. These species hybridize readily in cultivation.

Ecological Isolation`

Two species occur in the same general area but are separated ecologically, occupying different habitats e.g., *Silene alba* on light soils in open places and *S. dioica* on heavy soils in shade or in areas of high rainfall. Then habitats overlap in some cases, especially on road side merges and hybridization is then common.

Seasonal Isolation

Two species occur in the same localities but flower at different seasons e.g., *Sarmbucus racemosa* and *S. nigra*, the latter flowering 7-8 weeks later than the former. Hybrids occur in exceptional seasons when the flowering period overlap.

Temporal Isolation

Two species flower during the same period but the pollen is released and / or the stigmas are recapture at different times of the day e.g., *Agrostis tenuis* and *A. stolonifera*, the former flowering in the after noon and the later in the morning. Pollen release at an unusual time of day due to abnormal weather conditions or pollen remaining on insects or air borne for long periods leads to hybridization.

Ethological Isolation

Related, interfertile species are pollinated by different sorts of pollen vectors, or on any particular forage an individual animal keeps to one species only. e.g., *Aquilegia formosa* pollinated by humming birds, and *A. pubescens*, pollinated by hawkmoths. Hybrids arise from the inconstancy of vectors, especially when flowers of their preferred species are rare.

Mechanical Isolation

Related, interfertile species have differently structured flowers which makes it difficult for pollen vectors to transfer pollen to the stigmas of flowers other than those of the same species from which it was obtained e.g., *Ophrys insectifera* and *O. apifera*. Hybridization can result from the attention of clumsy insects often those which donot usually visit either of the species concerned.

Breeding Behaviour Isolation

Species which are predominantly inbreeding or apomictic hybridize for less frequently than species which are self incompatible. E.g., *Festuca rubra* x *Vulpia fasciculate*, the latter parent being semi-cleistogamous and largely self fertilizing. Hybrids with this species as the female parent are easily made artificially.

Gametophytic Isolation

Cross pollination takes place, but the pollen tube fails to germinate or to reach and penetrate the embryo sac of the female parent. This is by far the commonest isolating mechanisms. If the stigmas of almost any plant are examined microscopically pollen grains from other species will be found adhering to it, yet not germinating. Many of these species are of course only remotely related to the recipient plant. When the relation is closer the pollen may germinate but fail to penetrate the embryo sac. Examples are found in cereals, *Lathyrus* and *Datura*. The mechanisms can be over come in some cases by stylar surgery.

Gametic Isolation

The pollen tube releases the male gametes into embryo sac but gametic fusion or endospermic fusion does not takes place. This has been studied in various crop plants and other genetically well known genera.

Seed Incompatibility

The zygote or immature embryo ceases development, so that mature seed is not formed e.g., *Primula elatior x P. veris*. This phenomenon is very well known to plant breeders. It is often, though not always associated with failure of the hybrids embryo, in which case it can often be over come by excising the embryo and culturing it on nutrient media.

Hybrid Inviability

Germination occurs, but the resultant F1 hybrid dies some time before its production of flowers. Example *Papaver dubium x P. rhoeus*. This can also sometimes be overcome by culture techniques.

Non Fitness of F1 Hybrids

The F1 hybrids reach maturity but are not successful competitors in the wild. Usually this relates to the unavailability of suitable ecological niches, which may be different from those of either parent, and it is often coupled with ecological isolation as in *Geum rivale x G. urbanum*.

F1 Hybrid Sterility

The F1 hybrids are fully viable but sterile and therefore do not contribute to future generations. Sterility may be manifested from early (e.g. *Senecio squalidus x S. vulgaris*). The isolating mechanism is often, though by no means always associated with poor chromosomal pairing at meiosis, especially when the parents have different chromosome numbers.

F2 Hybrid Inviability (Or) Sterility

In these situations the disability is delayed until the F2 generation or even later. The F1 hybrids *Festuca rubra x Vulpia fasciculate* is largely (over 99%) sterile but a few F2 plants have been obtained. These are weak and do not flower.

Non Fitness of F2 Hybrids

This is again the case of a disability delayed until a later generation. This isolating mechanism is exemplified by those many situations in which two species are fully interfertile and produce fully fertile hybrids yet the two parents remain recognizable as distinct entities are often both more common than the hybrids. For various reasons the back crosses F2 and later generations are less successful in nature than the parental species. After a few generations a multitude of hybrid offspring might have been produced, demanding a multitude of ecological niches. The likelihood that all these requirements will be met is not high.

13

Breeding Systems

Objectives

i) To understand the series of mechanisms promoting cross fertilization and genetic system for evolving self incompatability.

ii) To understand the selective advantage in promoting the gene recombination via self incompatability.

iii) To understand that population contain a large store of variability hidden in the recessive genes in heterozygous condition.

iv) To understand the main breeding systems range from mechanical devices of the flower to prevent self pollination.

Introduction

An out breeding system is one in which sexual reproduction involves the mating and union of gametes of different individuals.

Plants can be predominantly out breeding, or some mixture of the two. In many flowering plants specific mechanisms have evolved that promote one of these systems. Out breeding, also called out crossing, allogamy or xenogamy is the transfer of gametes from one individual to another, genetically different individual.

In plants, out breeding may be obligate or facultative. The necessity or tendency for out breeding may be reinforced by dioecism, self incompatibility, heterostyly and other mechanism. Out breeding is usually by the wind, insects, or other pollinating agents. Out breeding may imply mating between individuals that are less closely related than would occur in random mating, as in parmictic populations in which each individual has the probability of mating with any other individual.

Dioecy

The probability of out breeding can be increased by a variety of mechanisms, in which individual plants have either male (sterminate) or female pistillate flowers ensures that out breeding will always occur. Many flowering plants exhibit a modified type of dioecy in which some individuals have flowers of one sex but others have bisexual (perfect) flowers. These include (gynodioecy (some individuals with pistillate flowers only others with perfect flowers) androdioecy (some individuals with staminate flowers only, others with perfect flowers) and trioecy (some individuals with staminate flowers, some with pistillate flowers only and some with perfect flowers). These alternate mechanisms may promote outcrossing but also allow for some inbreeding ensuring that at least some seed will be set.

Another out crossing mechanism is result of differences a timing of maturation of male and female floral parts a feature known, as dichogamy, of which there are two general type, protoandry is the precocious development of androecium as occurs e.g. in many members of Apiaceae, Asteraceae and Companulaceae. In Protandraceae species the pollen matures and is released prior to maturation and respectively of the gynoecium.

Protogyny is precocious development of the gynoecium as occurs e.g., in some chenopodiaceae. Both protandry and protogyny promotes out crossing when flowers of different individuals mature at slightly different rates. Thus, the pollen from one flower will not normally pollinate that same flower, but pollinate a different flower in which the gynoecium is receptive.

In protandrous and protogynous species, out crossing is ensured only if the flowers from a given individual mature at the same time. In reality, most of these species have flowers aggregate together into infloresences in which range of developmental stages may be present. In any case, at least some out crossing may occur and pollination within a single flower is normally prevented. However, if the pollen is not removed from a flower, it may in some cases pollinate the gynoecium of that same flower. This provides a fail safe mechanisms for producing seeds even in times or environments where pollinations are lacking.

Hercogamy

Out crossing has also been enhanced by evolutionary as in floral structure, particularly the spatial separation of anthers and stigmas, a phenomenon known as Hercogamy, also spelled herkogamy. (However, hercogamy may also function to prevent interference of male and female functions in the flower, by physically separating them). One type of hercogamy is heterostyl in which the relative lengths or heights of stigmas versus anthers is relatively constant. In so called distylous species, two floral morphologies occur per flower is likely to have pollen deposited on its body in a location but would effect pollination of thrum flower rather than another pin flower and vice versa

This increases the probability of pollination between flowers rather within flowers. If individuals tend to have one floral type or the other out crossing would be ensured. A rarer, more complex situation occurs in species that one tristylous, with heights of styles and stamens. The principle for cross pollination is the same.

Another type of hercogamy is enantiostyly (or

enantiomorphy) the curvature of the style to either the left or the right (as viewed from the front) defining left handed and right handed flowers. This style curvature usually corresponds with a curvature of at least one stamen to the side opposite the style. As with heterostyly, enantiostyly results in the preferential deposition of pollen on one side of, say, an insect pollinators body. For example, an insect visiting a left handed flower would tend to got pollen deposited on the right side of its body. Thus, the insect would more likely pollinate a right handed flower as opposed to another left handed flower. If an individual plant is relatively constant as floral handedness enantiostyly will greatly promote the probability of outcrossing.

Yet another type of hercogamy involves movement of floral parts. One type of movement hercogamy is the rapid closure of the stigmas upon their being touched by a potential animal pollinator (eg. mimirlus, phyromaceae). If the stigmas first receive pollen from a pollinator, their rapid closure can physically prevent pollen from the same flower being transferred to the stigmatic region, effectively preventing intrafloral self pollination. Another type of movement hercogamy involves trigger mechanisms (e.g. *Kalmia, stylidium*) in which an insect pollination triggers the sudden movement of one or more dusting the insect with pollen at the point of contact. This pollen is then at a position to be more effectively transmitted to the stigma of another flower.

Finally out crossing can be promoted by genetically determined self-incompatibility mechanisms. Self incompatibility refers to the inability for fertilization to occur between gametes derived from an individual genotype. Because this is genetically determined the incompatibility operates both within a single flower and between flowers of an individual.

Cleistogamous

The occurance of dimorphic flowers- Cleistogamous and Chasmogamous has been described for several members of the Commelinaceae. Detailed studies have been made on *Commelina forskalaci* (Maheswari & Balder, 1958). In this Indian monsoon herb the aerial flowers are blue and chasmogamous while underground flowers are white and cleistogamous. The production and development of the latter are influenced by light moisture they develop in greater abundance under comparatively dry conditions while in water logged soil few are produced. Cleistogamous flowers become chasmogamous on exposure to light.

The taxonomic importance of cleistogamy is two fold. Firstly in practical terms there is the danger of misinterpreting the taxonomic value of these cleistogamous forms. It has led, for example to the description of new leguminous genus (*Martia*) based on the cleistogamous morphism of a well known species of *Clittoria* (Van steenis, 1948). This is an extreme example, but failure to understand the positionary result in the recognition of varieties or species based on cleistogamous variants that are environmentally controlled. In evolutionary terms cleistogamy is important because of its effect on the variation pattern of the population concerned as a result of increased homozygosity. This can lead to a lack of genetic variability which may be advantageous in a stable environment, but is of course highly disadvantageous in a changing environment since the population will lack the capacity for change to meet the new conditions.

Mechanisms Favouring out Breeding

Many mechanisms favoring and varied. Some are found in habitual outbreeders, others is probably the majority of plants which show some degree of outcrossing, such plants may be monoclinous with stamens and carpels in the same flower (e.g. *Papaver, Hypericum, Althaa* etc.) or diclinous with unisexual flowers- either monoecious or dioecious.

The main out breeding systems range from mechanical devices of flower to prevent self-pollination (Synanthesis) and temporal feature having a similar effect, to dioecium and self incompatability.

Complete separation of the sexes (dicliny), naturally promotes outbreeding. The simplest case is monoecium, where the different sexes are borne on the same plant. This doesnot guarantee outbreeding but rather favours it. According to Baker it is wasteful of gametes if it is not combined with some feature to prevent selfing. Such device is found in *Zea mays* where flowers of different sex open at different times, thus making the outbreeding system more effective. Variations of the monoecious system are polygamy: andromonoecious where male and hermaphrodite flowers co-exist on the same plant. In *Parietaria diffusa* there is complex within a species: in *Mercurialis perennis*, is normally dioecious species, rare monoecious individuals occurs. Plant bearing these different flower forms can mislead the taxonomist when making his specific delimitations, especially in floras which have not been intensively studied. Van steenis cites the example of *Bauhinia castata* which is a female variant of *B. malabarica* and notes that the correlation of specimens representing different sex forms may cause difficulties.

Dioecium

Dioecium, where the sexes are separate on different plants, is the most drastic system of outbreeding, since not only is outcrossing obligate but it involves the very considerable risk that fertilization may not be realized at all (Baker, 1959). There is, for example, a strong possibility that in any locality the neighbouring plants in the population will be of the same sex. Furthermore, only about half the plants will be pistillate and therefore potentially capable of fertilization. Dioecism like monoecism may be incomplete.

Gynodioecium

Some species are polygamous (gynodioecious) where hermaphrodite and female individuals occur, they may also occur on the same plant. Gynodioecism is an outbreeding mechanism, since the hermaphrodite flowers may be selfed but the female flowers must be out crossed, the different mating types contributing unequal numbers of genes to the next generation (Lewis and Crowe, 1956).

Gynodioecism is frequent in many genera of the Labiatae such as *Oreganum, Salvia, Mentha, Calamintha, Glecoma,* Caryophyllaceae and Saxifragaceae. Detailed studies on the gynodioecious breeding system of species of *Primelea* (Thymeleaceae) have been made by Burrows (1960). In New Zealand species studied a high proportion of females was found in populations which ensured out crossing a low fruit set was found in the female flowers. Widespread hybridization in the genus is attributed by Burrows to the gynodioecious condition in many species.

Such sexual polymorphism may cause difficulties for taxonomist since the morphological differences in the flowers may be quite marked and if the differences are not understood, may result in the plants bearing them being described as separate species. In the Labiatae the female sex flowers are typically smaller than the hermaphrodite flowers, this difference may be mistaken for simple size variation in several cases the sexual nature of the polymorphism has not been appreciated. This was the case, for instance in the genus *Ziziphora* until recent investigations showed that sexual differentiation is quite common. There the length of the corolla tube, its degree of exsertion from the calyx and the degree of exsertion of the stamens depend on the sexuality of the plant (Hedge, 1961a). The interesting point in this case is that corolla tube length has been used by previous taxonomist as a key character in separating species of *Ziziphora.*

Gynodioecism should not be confused with the rare mutants. Causing male sterility in many hermaphrodite species. This is manifested in the abortion of pollen, but has little effect on outbreeding system of a population.

Within a genus both monoclinous and dioecious species may occur, for example in *Mussandra* where *M. tristigmatica*, , *M. cheppii* are functionally dioecious while *M. erythrophylla* is heterostylous. In this case the floral structural differences associated with different breeding systems provide useful taxonomic character although they have been overlooked (Baker, 1959a).

Dichogamy

In hermaphrodite plants differential maturation times may operate. This is known as dichogamy when the stigmas and anthers mature at different times. It exist in two forms protandry when the pollen is released before the stigmas of the same flower are receptive; and protogyny when the stigmas are receptive before pollen from the same flower is released. Both of these mechanisms normally result in cross pollination. The difference in the time of maturity between the pollen and stigma may be quite short, indeed dichogamy may be only partial with some overlap in maturation times and this may refer to individual flowers to individual plants. There is often some 'imperfection' of this nature in pollination devices maintaining some sort of balance between selfing and crossing. In some species such as *Zea mays* both protogyny and protoandry may occur, similarly some races of *Erodium cicutarium* are protogynous while others are protandrous. Dichogamous mechanisms are known in many families and are frequent for instance in the Gramineae, Juncaceae and Compositae.

Apart from dichogamy very numerous floral mechanisms and ethological factors favours out breeding, by adapting flowers to specific pollen vectors.

Geitnogamy

A factor which tends to be overlooked in out breeding mechanism is geitnogamy- the autogamous pollination that results from the transference of pollen from one flower to another on the same individual. It occurs, for example in *Eupatorium cannabinum* (compositae) between the flowers in the same inflorescence and in various members of the Umbelliferae such as *Chaerophyllum* and must be quite common in many other groups, although there is little information as to its effectiveness. It is one of the ways by which some out breeding mechanisms may be circumvented (Baker, 1959).

Heterostyly and Self Incompatibility Mechanism

Incompatibility systems favour out breeding. They depend, on the genotypic differences but they operate in hermaphrodite organisms which prevents the formation of zygotes and self mating (Lewis, 1954). It is wide spread amongst Angiosperms and is known from more than 3000 species in some twenty families and probably many more examples remain to be discovered. Few species have been well analyzed, largely because of the labour involved.

There are different degrees and types of incompat·bility. A distinction is made between self-incompatibility which operates between the pollen and style of the same plant (i.e., prevents the self-fertilization of hermaphrodites) and cross-incompatibilty which operates between the pollen and style different individuals, races.

Three main self-incompatibility are found in Angiosperms. The incompatibility reaction may be between pollen grain and stigma pollen tube and style, or between the male and female generative nuclei in the embryosac.

Heteromorhpic (Primulaceae, Lythraceae, Linaceae, Rubiaceae)

There are two or three morphologically distinct mating groups found here long the short style flowers in distyly, long, medium and short style flowers in tristyly. The pollen reaction is sporophytically controlled the plant control the characteristics of the pollen by means of morphological differentiation of the flower (Darlington, Mather, 1949) Genetic control is by one gene with two alleles S and s in distyly or by two genes each with two alleles Ss and Mm in tristyly.

Homomorphic

Homomorphic, gametomorphic (Solanaceae, Rosaceae, Leguminoceae, Scropulariaceae). Here all the mating groups are indistinguishable morphologically. The pollen reaction is gametophytically controlled. Genetic control is by one gene with numerous alleles S 1 2 3 .....n) multiple allelomorphic incompatibility.

Homomorphic sporophytic (Compositae, Cruciferae, Rubiaceae): This system combines features of the other two systems- genetic control is by one gene with many alleles but the pollen reaction is under sporophytic control.

Detailed discussion of self incompatibility are given by Bateman (1952) Lewis (1954 a,b) and Pandey (1960) whitehouse (1950) has produced evidence to suggest that multiple - allelomorph incompatibility of pollen and carpel tissue may well have been primary cause of evolution of the closed Carpel and of the success which the Angiosperms achieved over their ancestor gymnosperms.

Heterostyly

In the heteromorphic system there are various accessory characters involved. In the classic case of *Primula* e.g. *P. sinensis*, in the pin and thrum flowers morphological features are, style length and stigma position, anther height and position, size and configuration of the stigmatic papillae.

Such features are controlled by the same gene complex as in the Plumbaginaceae where the incompatibility system is allied to that of most heterostylous taxa, here there is a striking pollen dimorphism. Plants with the different kinds of pollen are found in populations of dimorphic species and those bearing each pollen type are self incompatible but cross compatible. In *Limonium vulgare* there is also stigma dimorphism with cob and pistillate stigmas corresponding to the ten kinds of pollen. There is also heterostyly with long cob styles and short appellate styles. This culminates an evolutionary sequence in the tribe Staticaceae, derived from a primitive condition in which there is monomorphism has arisen in *Armeria* and *Limonium* this indicates self incompatibility, but the plants are still capable of some degree of outbreeding (Baker 1953).

In *Linum grandiflorum* only the style differs in length. It is shorter or equal in length to the stamens. Tristyly is known in *Lythrum, Oxalis, Narcissus.* The mechanism in the latter is not strictly speaking heterostyly, however since the variation is not apparently related to the breeding group. It is known that in *Narcissus cernuus* only single positive may be found in a given area.

In *Pimula vulgaris*, Crosby (1949) has studied the change over from out breeding to inbreeding associated with a change from heterostyly to homostyly. The significance of this example is that it can be studied in the act it has probably occurred and been completed in other species of *primula*. Monomorphic species of the sections *Candelabra* and *Farinosae* are found near the edges of the range distribution suggesting their derivation from the main body of heteromorphic species, and monomorphic forms of normally dimorphic species of the sect. *Farinosae* have also been found at the margin of the range of these species.

According to Baker (1959,1961) all these system promoting outbreeding are reversible, permitting outbreeding. Thus morphological changes may take place in the flowers, self incompability system, heterostyly give rise to homostyly (as in the example of *Primula* cited above). The reverse evolution from inbreeding to outbreeding is apparently much rarer. This flexibility is of great evolutionary advantage permitting adaptation to changing conditions apomictic groups on the other hand are correspondingly limited.

Evolutionary Significance of Breeding

Darwin is of the opinion that plants should be cross pollinated by studying the pollination mechanisms of orchids. Darwin demonstrated that their purpose was the fertilization of the flowers by pollen brought by insect from other plants. Through cross pollination, the plants produce vigour to maintain fertility. In outbreeding the gene recombination and gene mutation bring about the variation upon which natural selection will be operating to bring out the adapted structures and organisms. This sort of selection may be called allaesthetic as in the case of the genus *Ophyrus*.

In out breeding population variability is partly held in latent form as reservoir of successive genes held in the heterozygous condition. Most of the variation is not adaptive and when brought out by recombination reduces the fitness of the population. This hidden variability is of considerable significance since it will produce an array of genotypes some of which may be adapted to meet new conditions which the species might come to face as in changing environments. According Lewis & Crowe (1958) when organisms are competing in a common environment with other organisms, the out breeding mechanisms are essential for survival by giving the advantages of gene recombination. Gause's principle (1934) indicated that two or more forms with identical requirements cannot coexist in the same environment on out breeding system may be retained for some reason be abandoned.

14

Recent Trends in Plant Systematics Anatomy, Palynology, Cytology, Phytochemistry, Numerical Taxonomy, Molecular Taxonomy

Objectives

- The plants have to be studied in entirety, to arrive at a more meaningful conclusion of judging the taxonomic status.
- Whether the application of anatomical data to the solution of taxonomic problems is essential.
- To assess whether Embryological data played a significant role in systematic considerations.
- To ascertain whether the Cytology often has bearing in systematics solutions.
- To assess the chemical data played a significant role in taxonomic problems or not.
- Evaluation of variation by numerical similarity among the groups of organisms.

Introduction

Systematics is one of the major subdivisions of biology broader in the base than genetics and biochemistry. It includes the service functions of identifying and classifying but comparative study of all the aspects of organisms. Contributions of taxonomy to the applied science includes medicine, Public health, agriculture conservation and management of resources. Most systems of classification of plants have been proposed on the basis of exomorphic characters placing emphasis on the flower characters which are conservative than the vegetative organs. In the twentieth century the technology advanced so much that newer evidences became available to the taxonomist from different sub-disciplines like anatomy, cytology, embryology, palynology, genetics, physiology, ecology, plant geography and phyto-chemistry etc. The numerical taxonomy and biochemical systematics have developed diverse and highly sophisticated set of procedures available for classificatory and discriminatory purposes than one might have predicted. Information processing side of taxonomy with electronic computer, scanning electron microscopy has reached very high levels of refinement and is virtually become every day tools for the study of the micro characters and that generate information and data along with ultrasonic featuring or sectioning, revolutionized systematic, palynology had been employed in taxonomy as a modern taxonomy and New taxonomy or experimental taxonomy. The plants have to be studied in entirety to arrive at a more plausible, reasonable and meaningful conclusion of judging the taxonomic status. The newer aspects of taxonomy like cytotaxonomy, chemo-taxonomy, numerical taxonomy, biosystematics etc have greatly clarified facts in the systems devised by classifical taxonomy.

The generated data from these sub-disciplines will be used to delimit, define and solving some taxonomic problems and proposed the classification based on synthetic theory but also revealed mechanism of evolution at various levels of taxonomic hierarchy.

Anatomy in Relation To Taxonomy

Anatomical characters play an increasingly important role in the formation of natural or phenetic groups. They have also been widely used to the elucidation of phylogenetic relationship. Source of wood anatomy data has often revealed possible affinities or has thrown new light on existing taxonomic arrangements some times indicating their weaknesses and possible faults.

The application of anatomical data to the solution of taxonomic problems is not new but as Metcalfe (1954) observes it is only within recent years, sufficient data have accumulated to draw firm conclusions.

Bailey (1951) wrote a paper on the use and abuse of anatomical data in classification and phylogeny. Bailey (1951) and Metcalfe and Chalk (1950) have cited numerical anatomical characters that can be made use in taxonomy.

In addition to this the information is also obtained from petiole anatomy, nodal anatomy, seedling anatomy and foliar anatomy.

By using light microscope the anatomical information can be obtained from the following anatomical aspects namely (1) epidermology, (2) floral anatomy, (3) wood anatomy. In recent years the anatomical information is procured from SEM and TEM studies. The application of these data is called ultrastructural systematics (Blair and Turner 1972).

Epidermal Characters

Stomata : Stace (1965) emphasized the role of epidermal characters in taxonomy. Vesque (1889) recognised four stomatal types, named after the families in which they were observed. Metcalfe and Chalk (1951) classified the stomata on the same plan with some modification in terminology part recognised 10 stomatal types mainly on the basis of their ontogeny Von Cotthem (1970) divided the stomata into 8 types. They are as follows :

(1) Anomocytic (Ranunculous type) irregular celled Stoma surrounded by a limited number of cells that are indistinguishable in size, shape or form from those of the remainder of epidermis. E.g., Ranunculaceae, Malvaceae, *Tridax procumbens, Boerhavia diffusa*

(2) Anisocytic (unequal celled, cruciferous type) stoma surrounded by three cells of which one is distinctly smaller than the two. e.g., Cruciferae, Solanaceae

(3) Paracytic : (Parallel celled rubiaceous type) Stoma accompanied on either side by one or more subsidiary cells parallel to the long axis of the pore and guard cells. e.g., Rubiaceae, Magnoliaceae

(4) Diacytic (Cross called, caryophyllous type) Stoma enclosed by a pair of subsidiary cells whose common wall is at right angles to the guard cells. e.g., Caryophyllaceae, Magnoliaceae.

(5) Tetracytic : Four subsidiary cells are present, two lateral and two terminal characteristic of many monocotyledons.

(6) Actinocytic : Stoma surrounded by a circle of radially elongated subsidiary cells.

(7) Cyclocytic : Stoma surrounded by four or more subsidiary cells which form a ring around each stomata.

(8) Hexacytic : Stoma accompanied by six subsidiary cells consisting of two lateral pairs parallel to the long axis of the pore and two polar (terminal) cells, the second lateral pair as long as the stomatal complex.

In addition to the above several other characters can be used in taxonomy presence of stomata on the upperside (epistomatic) or lower side (hypostomatic) or on both sides (amphistomatic) are also useful in taxonomy. To give one example, the presence or absence of stomata on the upper surface of a leaf is a good diagnostic feature in separation of *Alnus subcordinata* and *A. orientalis* of Betulaceae which are otherwise difficult to distinguish. Similarly the shape of the guard cells are also useful in plant classification.

For example stomata with bean shaped guard cells are met with in Dicotyledons while stoma with dumbel shaped guard cells are seen in Poaceae.

Based on ontogeny the angiosperm stoma has been divided by Pant (1965) into three types.

Perigenous : Stoma in which subsidiary cells do not have a common ongin with guard cells, but one formed by cells lying around the meristemoid (initials) as the guard cell initials.

Mesogenous: Stoma in which the subsidiary cells are produced from the same meristomoid (initials) as the guard cell initials.

Mesoperigenous : Stoma in which at least one of the subsidiary cells has a common origin with the guard cells, but others do not.

Development of stomata is useful in plant classification. Ramayya (1977) has emphasized the importance stomatal structures in taxonomy Ramayya and Rajagopal (1968, 1971) prepared a key to genera of *Portulacaceae* and *Aizoaceae* based on epidermal characters.

The stoma in the family Acanthaceae are diacytic where in closely related Scropulariaceae the stoma are anomocytic. The stoma in the family combretaceae are paracytic and stoma in the family strephonematoideae anomocytic in the subfamily combretoideae except the tribe Laguncularìaceae which they are cyclocylic. Nevertheless the stomatal characters not so reliable and the different types have obviously each originated on many different occassions and at times they all occur on the same plant. In two general *Streptocarpus*, and *Saintpaulia* belong the *Gesneriaceae* in which the Cotyledons bear anomocytic stomata and mature leaves anisocytic stomata. Sahasrabudhe and Stace, (1974). Pant and Kidwai, (1964) reported four types of stomata viz anomocytic, anisocytic diacytic and paracytic on the leaves of *Lippia nodiflora* of

Verbenaceae. The four above mentioned stomatal types were recorded in *Alternanthera pungens* of family Amaranthaceae.

Inamdar (1968) after studying the stomatal of *Nyctanthes* suggested that this genus should be kept in Oleaceae. The stomata are Tetra mesoperigenous in *Nyctanthes* as well as Oleaceae rarely aperigenous while verbenaceae shows diameso genous stomata. So Inanmdar (1968) indicated that *Nyctanthes* should be placed in Oleaceae instead of Verbanaceae.

Bremkamp (1953) felt that the genus *Elytaria* should be removed from Acanthaceae and should be placed in scorpulariaceae. Paliwal (1966) made intensive study of the phylogeny of stomata to ascertain whether this is correct or not. He observed that the stomata in *Elytaria* and *Acanthaceae* are diacytic and development is syndetochelic, but in *Scropulariaceaei* the stomata are anomocytic and development is Haplochellic. So Paliwala (1966) retained the genus Elytaria in the family *Acanthaceae.*

Trichomes

Trichomes are epidermal out growth there is such a wide range of trichomes types that classification of them constituted considerable problems. Trichomes are of great systematic significance and often even common types are used for diagnostic purposes in association with other characters. The recent trichome classification is that devised by Theobald *et al* (1979). Schulz (1936) used the type of hair as the major criterion in the subdivision of the family into Cruciferae into tribes and genera. Ramayya (1962, 1969) made a study of the trichomes of compositae and presented key to the genera. He considered the Heliantheae as primitive tribe based on trichome characters. Rollins (1944, 1945, 1946, 1949) made a study of the tricomes of Parthenium *argentatum* and *P.incanum* and their natural hybrids. He observed T-shaped trichomes in *P.argentatum* and long whip like thread types in *P.incanum* whereas intermediate types of trichomes are observed in the hybrids between these species.

Stace (1965, 1969, 1973) recorded that the trichome

anatomy is of immense significance in classification at all levels in the family Combretaceae from the circumscription of the family down to the separation of the species and even varieties virtually led to an improved tribal classification within the family and an improved sub-generic and sectional classification within the largest genus *Combretumn*. Theobald *et al* (1979) devised the most recent classification that is practical and straight forward.

Epidermal cells

The epidemis can provide a host of features and examples of their use abundantly occur in the literature. Epidermis include trichomes, stomata the Venation system other special structures, provide many charaters of taxonomic value. These include their shape, size, orientation, anticlinal wall undulation and periclinal wall curvature. The epidermal cell modification occurs also on the leaf margin and when silicified, calcified, suberized, gelatinised and contain various types of crystals. Silicified and suberized epidermal cells are of major taxonomic importance in the Poaceae. Linsbaur (1930) Borill (1961) has investigated the taxonomic value of epidermal papillae which are found in one of the mediterranean areas consist of *Dactylis glomerata subsp.hispanica*. According Ahmad (1974) the epidermal characters of different species of *Hemigraphis* and *Strobilanthus* are adequately pronounced as to justify their use in making distinctions at specific level. Ahmed (1976) corroborated the treatment of *Astercanthus longifolia* as *Hygrophila auriculata* by Heine (1962) on the basis of his study of foliar epidermis.

According to Carlquist (1961) the leaves of Angiosperms show great diversity. The nature of epidermis, its thickness type of mesophyll, sclerenehyma patterns, sclereids, Crystals, venation patterns etc are considered useful in classification. Duval Jouve (1875) recognised importance of the foliar anatomy in the systematics of grasses. Poaceae have two types of internal structures in leaves as stated by Avdulov (1931). In the Festucoid grasses the mesostoma around the stele is layered and mesophyll cells are straight while in panicoid grasses mesostoma may or not be present and the mesophyll cells are arranged radially.

Ayensu (1974) observed sclerenehyma patterns in the ***Velloziaceae*** that pattern can be used in separation of genera *Vellozia* and *Barbasenia*. According to Govindarajulu (1969) based on the internal structure, the genera in the Cyperaceae can be segregated.

Calcium oxalate crystals, and foliar crystals are found to be useful in classification and in taxonomic treatment. Succulent and arid plants posses laticifers cells or series of fused cells containing latex which belong to Arid regions. They vary widely in their structure and latex in their composition. Occurrence of latex and its absence has been of some taxonomic value. According De Bary, (1884) certain species in Aroideae lack laticifers or related structures while others have longitudinal rows of elongated cylindrical sac like cells. *Fumriaceae* which lack laticiferous (Sperlich, 1939) should be considered distinct from other related family *Papavaraceae* having a well developed latex system as opined by some other authors. It is one of the reasons for regarding these families as distinct from the rest of the families in Englier's *Rhoedales.*

Petiole Anatomy

Vascular system of dicotyledons distal end of a petiole is of considerable taxonomic importance and Metcalfe and Chalk (1950) illustrated several types of Vascular structures. Many taxa have special type of Vascular structures e.g. *Anemone vitifolia*. Two distinct vascular bundles arrangements were observed. The first with 1 - 2 median arcs, noted in *Phlomis* section *phlomis* and the species of *Eremostachys*. The second type, characterised by the occurrence of numerous separate bundles, often in a complete cylinder occurs in *Phlomis* section of *Phlomis* whereas two medullary arc bundles present in *Eremostachys*. This confirms that within *Phlomis* two distinct groups can be recognised and that *Phlomis* and *Eremostachys* are closely related.

Vascular structures of petiole is of much diagnostic significance in the genera *Phlomis* and *Eramostachys* of Labiatae Azizian and Cutler, (1982) and Petiole has separate vascular bundles occur in complete cylinder.

The taxonomic usefulness of petiole vascularization patterns has become increasingy apparent in recent years (Radford *et al* 1974). A classification of nodal types in megaphyllous plants was given by Sinnot (1914). In general three major types of nodal anatomy was proposed in Dicotyledons. (1) unilacunar (2) trilocunar and multilacunar. Trilocunar nodes occur in majority of Dicotyledons. In the primitive orders such as *Magnoliales, Piperales, Trochodendrales* multilacunar nodes occur as well as in advanced orders such as umbellales, and Asterales, while in Laurales, *caryophyllales, Ericales, Diapensiales, Ebenales primulales Myrtales* and a majority of families of Asteridae the unilacunar nodes occur.

In herbaceous members the stem anatomy has been studied where it has proved to be of some diagnostic value. Little strengthening tissue that occur in the stems of *Ranunculus repens* which are herbaceous members that is correlated with the creeping habit of the plant where as in the closely related species *R.acris* interfascicular tissue is Sclerified and so the erect habit is possible.

In the subgenus *Genuini* of *Juncus* the anatomical characters of stem and foliar bract can be used to distinguish British species. Further on the anatomical grounds it is possible to identify parents of several hybrids. In the anatomy of stems of *Dioscorea* considerable variation was observed which can be successfully used in the delimitation of species that are not easily separable on exomorphic grounds. *Dioscorea rotunda* and *D.cayennensis* which are closely related. Ayensu (1970) observed distinguishable anatomical characters that are useful in separating the two apart.

During the last five decades plant anatomist, notably Bailey and his associates, have established a number of lines of evolutionary specialization in true structure of stele. Excellent summarises of this work are given by Bailey (1957),

Tippo (In Gunderson, 1950). Carlquist (1961) and Eames (1961). The vascular bundles are arranged in a ring in dicotyledons whereas in monocotyledons the vascular bundles are scattered. Benthan and Hooker (1862 - 83) placed Basellaceae as a sub family in Chenopodiaceae. Anatomically Basellaceae are characterised by a single ring of vascular stands with intraxylary phloem in the larger bundles. On the other hand Chenopodiaceae are characterised by the development of polycyclic anomalous secondary vascular bundles.

Bentham and Hooker (1873-1876) split the *Gentianaceae* into two families - Menyanthoideae and Gentianoideae. But Hutchinson (1969) Cronquist (1981) and Takhtajan (1980) raised them to independent families. Vascular bundles in Menyanthaceae are Collateral whereas Gentianaceae show bicollateral vascular bundles.

Vessels evolved by the modification of tracheids and the most primitive vessels have members closely resembling tracheids. Trachieds cannot be derived from vessel.

Primitive type of vessels compared with the most specialised types

Primitive	*Specialised*
Slender in relation to length	Comparatively broad in relation to length
Gradually tapered extensively overlapping ends (in tang L.S.)	Truncated ends
Thin walled	Thick walled
Angular in T.S.	Oval or rounded in T.S.
Multiperforation in overlapping ends	Single perforation at end of each cell
Intervascular pitting scalariform	Intervascular pitting alternating multiseriate.

The primitive type of Angiosperm vessels differs from the tracheids only in the dissolution of the pit membranes in a number of the pairs of boardered pits at maturity. Such multiperforate or scalariform perforations preceded those with simple perforations.

Among those vessels with scalariform perforations plates with many bordered perforations and angular walls in transactional view would be primitive. Thus there was a gradual reduction in number of the perforations, and a fusion of the bars between them to produce a wide single opening. It would be noted that modification in the perforations of the vessels walls is ontogenic. The evolution of the vessel element has occurred along with the following lines.

i) Decrease in the vessel element corresponding to advancement

ii) Alteration of end walls from oblique nearly transverse angle

iii) Loss of borders and the decrease of bars on the perforation plates

iv) Lateral walls of vessels show evolutionary series in pitting type from scalariform to transitional to opposite to alternate

v) transition occur from vessels with angular in outlines to vessels with nearly circular outline.

There has been a similar sequence of specialization of pittings in the side walls intertracheary walls from scalariform boardered to transitional, opposite, and finally alternate.

Wood anatomy in relation to taxonomy

The data from wood anatomy are of considerable taxonomic and phylogenetic significance. This branch of study has provided additional evidences which have proved invaluable in the solution of taxonomic problems. Wood anatomy can be used at all taxonomic levels. Both at and above the family level it has been successfully used along with other lines of evidence to establish the systematic position of the primitive vessels Angiospermic families Winteraceae (comprised seven genera *Drimys, Pseudowintera, Bubbia, Belliolum, Exospermum, Zygogynum* and *Tetrathalamus*) *Trochodendraceae, Tetracentraceae, Amborellaceae* and *Chloranthaceae*. The Englerian group of primitive Angiosperms, i.e. Amentiferea cannot be considered primitive since they have specialized wood contrary to the opinion of phylogenetist (Tippo, 1938) that the *Amentifrae* were primitive angiosperms. Taking into account of anatomical evidence from *Paeonia* and *Austrobaileya* that supports the case of considering the remarkable genera as the type of independent families (Bailey and Swamy, 1949).

At generic and below generic level, the most useful contribution of wood anatomy is to support for placing taxa of uncertain affinity. With the help of anatomical evidence in Quercus (Fagaceae) different sub genera and sections can be delimited with the evidence of wood structure the placement of the genus Myristica and related genera close to Lauraceae is supported (Eames, 1961). Whereas based on the wood anatomy Myristicceae are considered unrelated to *Annonaceae* and *Eupatomaceae*. Similarly based not on wood structure evidence, placing of *Calycanthaceae* in or near Rosales or *Myrtales* is supported sieve cells also provide taxonomic evidence e.g. Unspecialised sieve cells occur in lower vascular plants. Pteridophytes and gymnosperms along with one Angiosperms *Austrobaileya scandens*, exibited unspecialised sieve cells in phloem.

Floral Anatomy and Taxonomy

For determination of status of high ranking groups such as genera and species the floral Anatomy has played very important role. The role of floral anatomy in relation to taxonomy has been discussed in some details by Eames (1935) Puri (1952, 1958, 1969) and Murty and Puri (1980) anatomical pattern.

The affinities of families based on floral anatomical patterns in Papaveraceae, Capparaceae, Brassiaceae and

Magnoliacea could be established. The families commonly exhibit parietal placentation and inverted placental bundles with regard to floral axis on inner side of secondary marginal strands with normal orientation. This fact also suggest the placement of Moringaceae in Rhoedales. Puri (1947, 1949) could able to establish affinities among Passifloraceae, Moringceae and Cucurbitaceae on floral anatomical basis.

Burt (1965) concluded that *Cyrtandromea* has no close affinity with members of Gesneriaceae and suggested its transfer to Scrophulariaceae with which it exhibits more resemblance in terms the sepal lateral traces that emerges independently in several lateral traces in the carpels, compound ventral strand as well as bilocular ovary (Singh and Jain 1978) have supported the treatment based on the above floral anatomical characters. Kale and pai (1979) suggested the removal of the genus *Trichopus* based on the floral anatomy from the family Dioscoreceae of Bentham and Hooker and erection of a unigeneric family Trichopodiaceae as has been previously suggested on anatomical evidence by Ayensu (1966, 1972).

Fruit wall and seed coat Anatomy

Fruit wall and seed coat anatomy played a very important role in taxonomy. The distinction between the two genera *Ammonia* and *Rotala* (Lythracae) have been become a taxonomic riddle for quite some time. The diagnostic feature was in rupturing fruit irregularly (fruits) in *Ammania* and valvular capsules in *Rotala*. This difference cannot be displayed in herbarium specimens. Van Leemjen (1971) showed that the fruit wall character show distinction between the two taxa. The orientation of the cells of the two main layers of pericarp are different in *Ammania* the cells of both layers are very similar whereas in *Rotala* the cells of the inner layer are linear strongly lignified and transverse to those of the outer. Two parallel views were expressed regarding the systematic position of the Asteraceous genus *Lagasea*, Cassini (1815) Lessing (1832) DeCandolle (1836) and Maxmuller (1954) included the genus

Lagascea in the tribe vernoniceae. Bentham (1873) Hoffmann (1894) and others placed *Lagasea* in the tribe Heliantheae. Morphological and chromosomal studies support the assignment of the genus Lagascea to Heliantheae. Based on the pericarp structures, Pullaiah (1981) advocated the retention of Lagasea in the tribe Heliantheae. Three different zones are distinguished in the pericarp. Between the middle and outer zones; characteristic Schizogenous cavities developed which get filled with a brown resinous and hardening substance . Such Schizogenous cavities are absent in vernoniceae seed coat morphology has given a valuable taxonomic information. For systematic purposes the seed coat anatomy has played important role.

Ultrastructures in Relation to Taxonomy

The information obtained by the Scanning Election Microscopy (SEM) and Transmission Election Microscopy (TEM) was extensively used in taxonomy. Nevertheless, only a few characters are used significantly in plant classification. Dilated Cisternae (DC) in the endoplasmic reticulum were first discovered by Bonnet and Newcomb (1965) in the roof cells of *Raphanus sativus* (raddish) and later by Favali and Gerola (1968) in the phloem parenchyma of foliar veins in *Brassica chinensis*. Jorgensen *et al* (1977) detected them in foliar veins of *Capparis cynophallophora*. Iverson and Flood (1970) found that such structures are of common occurrence in the Cucriferae and these onganells are typical of Brassicaceae and Capparaceae.

About two decades ago Behnke (1972. 1977. 1981) in a series of observations he could able to demonstrate the ultrastructural details of sieve element plastids which can be used in plant Taxonomy. In 1971 and 1972 Behnke wrote on the taxonomic importance of sieve tube plastids. Behnke (1977, 1981) studied more than 1500 species from 3800 families with regard to this characters. Behnke (1975) reported the leucoplasts in phloem sieve tube elements and can be classified into two major categories, with sub groups, the S-type or starch accumulators and P-type, protein accumulators. P-type accumulators may also have some starch associated with the protein crystals but the S-type do not accumulate protein. There are however certain groups of plants *Rafflesiaceae, Crassulaceae* and some species of *Ulmaceae, Moraceae* and *Urticaceae* in which the plastids have neither starch nor protein accumulation and they are called S-type. S-type sieve element plastids. The sub class Ranunculidae, Hamamelididae and Dilleniidae have without exception S-type sieve element plastids. While in the *Rosidae* a great majority of the orders have this type. Half of the members of Magnolidae have S-type and half P-type elements. Behnke (1977) there are about 65 per cent of these have such plastids in the sieve tube elements.

In Caryophyllidae majority of the members show P-type sieve element plastids. P-type plastids are further differentiated on the number and shape of the crystalloids and the nature of the filaments surrounding them. PI sub type sieve element plastids are present in Magnoliales / Laurales, PII sub type in monocotyledons, PIII subtype in centrospermae and PIV in Fables.

Behnke (1974) carried out the investigations on the order Rhamnales and found that the family Rhamnaceae has s-type while vitaceae and Leeaceae had P-type sieve element plastids. The heterogeniety of the Magnolidae related to its plastid types is attributed to its basal position in the evolution of flowering plants (Behnke, 1975) P-type are considered to be ancestral and S-type exhibits derived condition by loss of protein. P-I subtype is considered as the most primitive among P-subtypes of flowering plants.

In explaining the inter-relationships and circumscribing the order caryophylloles (Centrospermae) Behnke (1975). PIII subtype sieve element plastids occur in all the families that are kept under this order. These include the following families. *Aizoaceae, Amaranthaceae, Basellaceae, Cactaceae, Chenopodiaceae, Didieriaceae, Nyctaginaceae, Phytolaccaceae, Portulacaceae, Caryophyllaceae* and *Molluginaceae*. In earlier classifications other families like *Polygonaceae viviniaceae, Bataceae, Gyrostemonaceae, Plumbaginaceae, Fouguieraceae, Frakeniaceae, Rhabdodendraceae* and *Theligoniaceae* have been kept in centrospermace but in recent classifications, these have been kept apart from centrospermae because they exhibit S-type sieve element plastids.

Embryology in Relation to Taxonomy

In the late nineteenth century, Hofmeister and Strausburger indicated the possibility of using embryological characters in taxonomy, by the 2nd quarter of 20th century embryological data has been actually used in taxonomy. On account of wider embryological similarities (embryological details among the species of a genus, genera of a family, families of an order) lack of detailed studies along with laborius laboratory work, the taxonomic unrealised values of embryology has not been fully realised. (Johri, 1984). A discerning application of embryological data, has solved many a taxonomic problem; at different levels of hierarchy from species to order and also suggested in assessing the affinities and tracing the phylogeny. In addition to other disciplines, embryology is bestowing disciplines and the taxonomic verdict should be based with other disciplines of botany for the sake of multiple correlations. According Maheswari (1950, 1963) Johri (1984) and Johri et al (1992) the following embryological parameters are usually employed in taxonomic considerations. The role of embryology in taxonomy have been summerised by many embryologists like Maheswari (1950), Johri (1963) Davis (1966), Poddubnaja Arnoldi (1976) Herr (1984) and more recently by Johri et al (1992) Pullaiah *et al* (2001). According to Maheswari(1950,1963) Johri (1984) and Johri et al. (1992) the following embryological parameters are usually employed in taxonomic considerations.

Embryological Features of Taxonomic Importance

1) **Anther** : Number of vertical rows of cells of archesporium, type of wall development (Basic, Monocotyledons, Dicotyledonous and reduced), the tepetum glandular amoeboid, number and ploidy of

nuclei in tapetal cells, successive or simultaneous cytokinesis, types of tetrad.

2) **Pollen grain** : Number and position of germpores, number of cells at the shedding stage.

3) **Ovule**: Type of ovule (Orthotropous, anatropous, compyltotropous, amphitropous or circinotropous) number of integuments, micropyle formed by one (inner to outer) or two integuments, nucellus crassinucellate, pseudocrassinucellate, presence or absence of endothelim, epistase hypostase, aril, caruncle, obturator etc.

4) **Archesporium**: Number or archesporial cells, extent of parietal tissue.

5) **Megasporogenesis and development of embryo sac** : Shape of the megaspore tetrad. Position of functional megaspore in the tetrad; embryosac development mono, bi-or tetrasporic number of nuclear divisions.

6) **Mature embryosac** : Number and organisation or cells in the embryosac.

7) **Fertilization**: Path of entry of pollen tube.

8) **Endosperm** : Nuclear, cellular or Helobial, presence or absence of haustoria, seeds albuminous or exalbuminous.

9) **Embryo** : Development type (Asterad, onagrad, chenopodial, solanad and caryophyllad), presence or absence of suspensor haustoria.

10) **Seed Coat** : Number of layers and their derivation (from one or two integuments).

11) **Special features** : Apomixis, Polyembryony.

Species level

Hooker (1888) Gamble (1901) and Cooke (1907) and others treated *Burmania colestis* as a species and variant of it as *B.colestis var pusilla*. The variety is lesser form than the former and possess pedicillate flowers, orbicular perianth parts with slightly broader apex while the species *colestis* is robust plant with sessile flowers, narrow perianth parts with round apex.

Earnet (1909), Earnest and Bernard (1912) studied the embryology of *Burmania Colestis*, Arekal and Ramaswamy (1973) made embryological investigations on the varietal form and compared the same with the previous observations which are summed up in the following table.

Burmania Colestis	*Burmania Var-pusilla*
1. Apomictic	Normal sexual cycle
2. Diploid non-functional pollen	Haploid functional pollen
3. Panthenogenesis present	No panthenogenesis
4. Poly embryony present	No polyembryony
5. Chalazal chamber of Helobial endosperm 2-nucleate	Chalazal chamber of helobial endosperm 4-6 nucleate

Based on the above embryological differences, Areckal and Rangaswamy (1973) indicated that elevation of the varietal form into an independent species. It is named as *Burmania pusilla* and placed on par with *B.colestis.*

Embryological solutions to the taxonomic problems

For delimitation of plant groups at all levels the embryological characters are best utilized. Double fertilization (Syngamy and triple fusion) which is seen in Angiosperms at the highest level and in no other division of plants it is found. Similarly nutritive tissue for the development of embryo is seen in Angiosperm and occur in gymnosperms. It is of pre-fertilization character. The dicotyledons differ in the cytokinesis of pollen mother cells after meiosis with monocotyledons. Simultaneous cytokinesis is observed in Dicotyledons whereas successive cytokinesis occur in Monocotyledons.

The subclasses are differentiated on embryological grounds within Dicotyledons and Monocotyledons. The Asteridae are characterised by unitegmic and tenunucellate ovules where as the caryophyllaceae are characterised by bitegmic and crassinucellate ovules.

Many a time the embryological data supplement other evidences in many cases. Various taxonomist are of opinion that *Polygonaceae* has been associated with the *Centrospermae* group. Polygonaceae are characterised by the orthotropous ovules where as centrospermaceae have either campylotropous or amphitropous ovules.

Embryological features offered solutions to some families

Podostomaceae: It is an absolutely tropical family, the family includes perennial aquatic herbs which grow attached to stones in running water. The family shows interesting embryological features. Pseudoembryosac formation due to the disentigration of nucellar cells below the embryosac is unique feature. A combination of this characters further mark out this family.

a) Occurrence of pollen grains in pairs.

b) Anatropous, bitegmic and tenuinucellate ovule with micropyle formed by the outer integument.

c) Bisporic embryo sac

d) Solanad type of embryogeny.

e) Prominent suspensor haustoria

f) Absence of antipodal cells (except in Dicraea)

g) Absence of triple fusion and consequently endosperm.

Onagraceae: The presence of *oenothera* type of embryo sac development is characteristic feature of only Onagraceae and it is found in no other family. In the megasporogenesis the four mega spores formed after meiosis, the chalazal three megaspores degenerate whereas the micropylar megaspore develops into embryo sac. The family is so characterised by the absence of antipodals. A *diploid endosperm* is very specific characteristic feature of Onagraceae.

Cyperaceae: In the family Cyperaceae the development of microspore from the pollen mother cell is quite unique feature. In all other Angiosperms the pollen mother cell divides by meiosis, resulting in four haploid nuclei, each of which forms functional microspore. A special mode of pollen development was observed in Cyperaceae family of the four nuclei formed after meiosis only one functions. The functional nucleus remains in the centre and the three nonfunctional nuclei are cut of on one side of the cells and they do not form pollen grains. The functional nucleus divides to form vegetative cell and a generative call. The non-functional nuclei eventually degenerates.

Loranthaceae: Two sub families Loronthoideae and Viscoideae of family Loranthaceae are quite distinct. Besides *floral structure* they differ in the mode of development of *embryosac, endosperm,* and *embryo.* In the Loranthoideae the viscid zone of first is situated outside the vascular bundles of floral tube, while it is located inside the vascular bundles in the viscoidea. On the basis of these differences it has been suggested that the two sub families be raised to the status of families.

Taxonomically disputed genera

Paeonia

In earlier system of classification *Paeonia* has been included in the family Ranunculaceae. Embryologists offered the solutions to the controversial genus *Paeonia* and regarding its systematic position. According to Hutchinson (1959) *Paeonia* has been removed from *Ranunculaceae* and placed in a separate family *Paeoniaceae.* In contrast with the rest of the genera of the *Rananculaceae* in *Paeonia* the generative cell is large and elongated, the embryosac is long and narrow, the seed coat is massive and germination is hypogeal. The embryogeny in *Paeonia* unique and is not found in other members of the Rananculaceae. Kumuzawa (1935) studied wood anatomy and affirmed the above conclusion. Wodehouse (1936) observed that *Paeonia* is different in pollen exine structure with the rest of the Rananculaceae members. Gregory (1941) noticed that the basic chromosome number of *Paeonia* is five in contrast to seven, eight, or nine in the Rananculaceae. Eames (1951) also pointed out that the floral anatomy of *Paeonia* is quite different. Based on embryological features, the separation of *Paeonia* from Ranunculaceae to a separate family was justified. The most peculiar embryological features of this genus is its embryogenity (Yakovlev and Yoffe, 1957; Murgai 1959, 1962) according to Yakovlev and Yoffe (1957) the zygote undergoes repeated nuclear divisions forming ceonocytic structure (Fig 14.1).

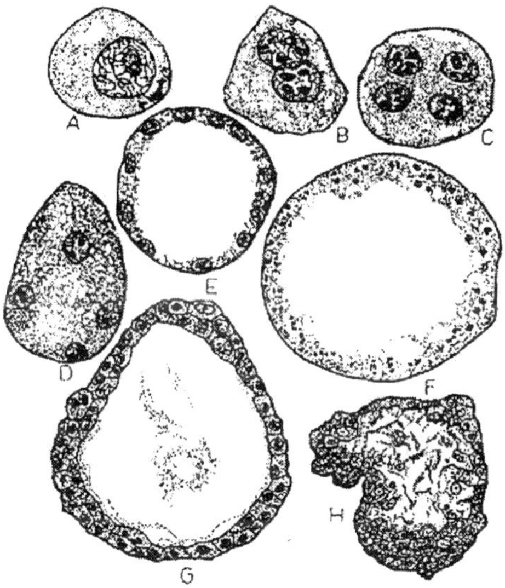

Fig 14.1 : Embryo Development in *Paeonia*

Trapa

The genus Trapa belongs to the family Onagraceae in Bentham and Hooker's system of classification. In Englarian system it was put in a separate family a treatment followed by most of the recent systematists. This view is further confirmed on the basis of the embryological evidences. Oeonothera type of embryosac is characteristic feature of Onagraceae while it is polygonum type in *Trapa*. A nuclear endosperm is present in the Onagraceae; such endosperm is lacking in *Trapa*.

Table 14.0 : Embryological differences between Trapa and Onagraceae

Feature	*Trapa*	*Onagraceae*
Pollen grain	Pyramidal with 3 much folded meridional crests	Bluntly traingular and basin shaped
Ovary	Semi-inferior bilocular with a single pendulous ovule in each chamber	Inferior, mostly trilocular with many ovules per chamber on an axile placentata
Embryosac	Polygonum type	Oenothera type
Endosperm	Absent	Present nuclear
Embryo	Solanad type	Onagrad type
Suspensor	Well developed suspensor haustorium	Short and inconspious
Cotyledons	One cotyledon extremely redcued	Cotyledons equal
Fruit	Large, one seeded drupe with prominent spines	Loculucidal capsule

Nyctanthes has been bestowed a place in Jasmenioideae along with *Jasminum* and *Menodora* by Bentham and Hooker (1862) Engler and Prantl (1931) has placed in monogeneric Nyctanthiodeae of Verbenaceae. Kundu (1966) suggested an independent familial status Nyctanthaceae to *Nyctanthes*. Kapil and Vani (1966) investigated the embryology of *Nyctanthes*. They compared the embryological features of *Nyctanthes* with those of Oleaceae and Verbenaceae. Their characters as follows:

In Nyctanthes nucellar archesporium is 1-2 celled whereas in Oleaceae it is one celled or 3 celled while in Verbenaceae it is 6-celled. Embryosac in *Nyctanthes* is monosporic and bisporic types, while in Oleaceae twin embryosac antipodals small and ephemerale are present. Monosporic and bisporic types antipodals small and epidermal where as in Verbenaceae monosporic type and antipodals large and pesistent with twin embryo sacs found. No twin embryosacs found.

Cellular endosperm 1st two divisions are transverse, chalazal endosperm haustoria absent in *Nyctanthes* whereas in Oleacea the same is the case like that of *Nyctanthes* while cellular endosperm first two divisions vertical or transverse chalazal endosperm haustoria present. Embryo is soland in *Nyctanthes* and Oleaceae while onagrad type in Verbenaceae. Seed coat is vascularised both in *Nyctanthes* and Oleaceae while it is non vascularised in Verbenaceae. Pollen grains are spheroidal tricolplate, sexine reticulate piloid, excrescenes present, Tenoinexious in both *Nyctanthes* and Oleaceae while peroblate prolate. 2-5 colpate sexine reticulate. No piloid excrescenes, sexine and nexine are of equal thickness. From the above embryological evidences it is evident that *Nyctanthes* shares with every embryological characters with Oleaceae and differs significantly from Verbenceae, so its retention in Oleaceae is more apt.

Table 14.1 : Embryological differences between Paeonia and Ranunculaceae

Feature	***Paeonia***	***Rananculaceae***
Stamens	Spirally arranged, centrifugal	Spirally arranged centripetal
Anther	Multilayered endothecium mostly 2 layered tapetum	One layered endothecium
Pollen	Reticulately pitted exine, large exine and elongate, generative cell.	Granular papillate smooth small generative cell
Female archesporium	Multicelled, many megaspore mother cells functions	Uni or multicelled, one cell functions
Antripodal cells	Persistent not polyploid	Persistent (ephemeral in Adonis) nuclei one or more than one, polyploid
Embryogeny	Unique, fre nuclear	Onagrad or rarely soland type
Seed	Arillate	Non arillate
Fruit	Follicle	Achene

The development embryo is onagrad type in all other Onagraceae, while it is solaned type in *Trapa*. Well developed suspensor and haustoria are observed in *Trapa*. While it is very short and inconspicuous in the *Onagraceae*. *Trapa* has very unequal cotyledons in contrast to equal colytedons in the Onagraceae.

Butomus : While redefining the Butamaceae, Pichon (1946) suggested that all genera of the Butamaceae except Butomus should be transferred to the *Alismataceae.* Cronquist (1968, 1981) and Takhtajan (1969, 1980) also retained only Butomus in the Butomaceae and transferred all other genera to the *Limnocharitaceae*. On embryological evidences also suggest Butomus stands apart from the other members of the family which justifies its segregation. While Butomas has polygonum type of embryosac the other genera are characterised by an Allium type of embryosac.

Different views have been expressed about the systematic position of Basellaceae a small family of herbaceous twining wines that occur chiefly in tropical America and Asia. Eichler(1878),Volkens(1893) and Lawrence(1951) include the family in Centrospermales- close to Portulucaceae. A similar allotment is offered by Bessey(1915) Gunderson(1950) Takhtajan(1980) and Cronquist(1981) although these systematist have adopted the alternative order name Caryophyllales. Hutchinson (1973) placed the family among group of advanced Caryophyllaceous families which he elevated to separate order, chenopodiales. Hooker (1880) includes it in chenopodiaceae as the sub family Bassellacea and Franz (1908) lowers it to a tribal rank, placing it in sub family Montioideae of portulacaceae.

Maheswari Devi and Pullaiah (1975) made a study of the life history of *Basella rubra* and summarised the embryological features of the family Basellaceae. They made a comparison of these features, together with other morphological characteristics, with those of the families included in centrospemales (Caryophyllales). The embryological features are compared in Table 14.2.

Table 14.2 : Embryology of Phytolaccaceae. Aizoceae, Basellaceae, Caryophyllaceae and Chenopodiaceae. (Adapted after Maheswari Devi and Pullaiah, 1975)

	Phytolaccaceae	Aizoaceae	Portulacaceae	Basellaceae	Caryophyllaceae	Chenopodiaceae
Anther	Tetrasporangiate	Tetrasporangiate	Tetrasporangiate	Tetrasporangiate	Bisporangiate	Tetrasporangiate
Wall	—	Basic type	Monocotyledonous	Monocotyledonous	Mono- or dicotyledonous	Monocotyledonous
Endothecium	Fibrous	Fibrous	Fibrous	Fibrous, multiseriate	Fibrous	Fibrous
Tapetum	Glandular, cells bi- or multinucleate	Glandular, cells bi- or four-nucleate	Gladular, cells bi- or multinucleate	Glandular, cells multinucleate	Glandular, cells bi-nucleate	Glandular, cells uni- or binucleate, occasionally amoeboid
Tetrads	Tetrahedral, iso-bilateral	Tetrahedral, isobilateral, decussate	Tetrahedral, isobilateral	Tetrahedral, isobilateral	Tetrahedral, decussate	Tetrahedral, isobilateral, decussate
Pollen grain	Two- (*Phytolacca*) or three-celled (*Rivina*)	3-celled	3-celled	3-celled	3-celled	3-celled
Ovule	Anacampylotropous, bitegmic, crassinucellate	Anacampylotropous, bitegmic, crassinucellate, funicular obturator develops	Anacampylotropous, bitegmic, crassinucellate	Ana- or campylotropous, bitegmic, crassinucellate	Campylotropous, bitegmic, crassinucellate, nucellar beak conspicuous	Campylotropous, bitegmic, crassinucellate, nucellar beak conspicuous
Micropyle	Formed by inner integument	Inner integument	Inner integument	Inner integument	Inner integument	Inner integument
Air-space between the integuments at their base	Present	Present	Present	Present	Present	Present
Archesporium	2- or 3-celled	Multicelled	2- or 3-celled	2- or 3-celled	Uni- or multicelled	2- or 3-celled
Embryo sac	Polygonum type	Polygonum, Endymion, Drusa, Adoxa or Penaea	Polygonum	Polygonum	Polygonum	Polygonum; Allium type in *Sueda maritima*
Antipodal cells	Persistent, secondary multiplication in *Rivina*	Ephemeral	Ephemeral	Ephemeral	Ephemeral supernumerary due to divisions	Ephemeral, sometimes
Endosperm	Nuclear type	Nuclear, chalazal haustorium in *Hereroa hesparantha*	Nuclear	Nuclear	Nuclear, in some taxa superficial extensions develop	Nuclear
Embryogeny	Caryophyllad type, massive suspensor	Solanad, occasional suspensor polyembryony	Onagrad or Caryophyllad, occasional polyembryony	Chenopodiad	Caryophyllad, basal cell of suspensor haustorial	Chenopodiad, suspensor massive
Seed	Small, exotestal, perispermous	Exotestal, and endotegmic, perispermous	Small, exotestal, albuminous, perispermous	Small, exotestal, perispermous	Small, exotestal, perispermous	Small, exotestal and endotegmic, perispermous

Embryologically, Basellaceae and Chenopodiaceae differ so apparently and inclusion of Basellaceae as a sub family of chenopodiaceae is untenable. Metcalfe and Chalk (1950) supported the above conclusion based on anatomical features.

In Basellaceae the stems contain a single ring of Vascular strands with intraxylary phloem in the larger bundles. Chenopodiaceae on the other hand are characterised by the development of polyclic anamolous secondary vascular bundles. Many embryological features shared by Basellaceae and Portulaceae supporting close resemblance as also suggested by Takhtajan (1980) Nevertheless to include, Basellaceae as a tribe in Portulaceae does not appear justified in view of certain differences.

All the embryological features exhibited by the Bassellaceae which are predominate in Caryophyllales so retention seems to be fully justified on embryological grounds.

Gentianaceae and Menyanthaceae

Bentham and Hooker (1873-1876) and Rendle (1962) Split the Gentrianaceae into two sub families (a) Menyanthoidea including Mashy plants and (b) Gentianoidea including all the others. Wettstein (1935) raised Menyanthoideae to a family rank. Which was accepted by Don (1938) and Lindsey (1936). This was supported by Vegetative as well as floral anatomy by Perrot, (1897) Lindsey, (1938) and embryology (Soueges; 1943; Maheswari Devi, 1962). The Gentianaceae should be retained in Gentianoideae to be better designated as Menyanthaceae now is justified.

The two sub families Menyonthoidea and Gentianthoidea exhibit the following morphological and embryological dissimilarities. Maheswari Devi (1962) advocated that these differences fully justify a family rank for the Menyonthoideae which would be better designated as Menyanthaceae. The name Gentranaceae should be retained by Gentianoideae.

Maheswari (1950, 1963) has listed twelve features judged to be of special interest in taxonomic considerations in delimiting larger plant groups.

The following embryological characters proved to be significance of taxonomic.

1. Number and arrangement of loculi in another, structure and thickening of endothecium and nature of tapetum has

proved.

2. Mode of quadripartition of the microspore mother cells
3. Development and organization of the pollen grain
4. Development and structure of the ovule.
5. Form and extent of nucelus
6. Origin and extent of the sporagenous tissue in the ovule.
7. Time of wall formation in the megasporonesis, arrangement of megaspore and development of embryosac.
8. Form and organisation of the mature embryosac
9. Fertilization - pathways of pollen tube and interval between pollination and fertilization
10. Type of endosperm and presence and absence of endosperm
11. Form and organisation of mature embryohaustoria.
12. Certain abnomalities of development such as parthenogenesis, apogamy, adventine embryony and polyembryony.

Table 14.3 : Differences between Menyanthoideae and Gentianoideae

Features	***Menyanthoideae***	***Gentianoideae***
Habit	Aquatic or marsly	Terrestrial
Morphology	Leaves alternate, corolla valvate or imbricate	Leaves opposite and decussate, corolla twisted
Anatomy	Vascular bundles collaterai, sieve tubes, absent in phloem	Vascular bundles bicollateral or amphicribal, sieve tubes present in phloem
Anther tapetum	Secretary type, cells bi- or multinucleate, nuclear fusion common	Secretary (Exacum) or amoeboid (Canscora) cells remain uninucleate
Placenta	Supressed parietal placenta	Well developed parietal placenta, some times the placenta fuse at the cntre to form a bilocular ovary
Endothelium	Present	Absent
Endosperm	Cellular	Nuclear
Embryogeny	Asterad type	Solanad type

Palynology

The term palynology was first used by Hyde and Williams in 1845.

The study of pollen grains and spores is known as Palynology. The use of palynological data for taxonomic solution of the problems is known as palyno-taxonomy. Based on palynological data and apertural patterns in particular the primitiveness and advancement among taxa can be traced.

There has been a great increase in comparative analytical studies of pollen morphology and wall structure. The study of pollen grain characters are of immense significance in evolution and phylogeny systematic and phylogenetic value have been obtained and these have been in agreement with conclusions drawn from other fields of study, particularly anatomy. The main characters of taxonomic value in pollen grains are (1) Number and position of furrows (2) number and position and complexity of the aperture and (3) the form of sculpturing of the exine. (4) Variation is also shown in size and general shape (5) Pollen association (6) pollen nuclear number.

Apertural patterns

On the basis of number, position and character of aperture (NPC). It is called NPC system of classification of pollen grains. Pollen grains are classified into 21 types based on the number; position and character of aperture. Majority of Angiospermous pollen grains fall into only two general types. Uniapertuate and triaperturate from which some less common types have been derived. Uniaperturate pollen grains have a single germinal furrow or pore situated at the proximal or distal position. The proximal form of aperture is found in Pteridophyte but in pteridosperms the position is reported to be either proximal or distal. Primitive angiosperms exhibit proximal aperture while the majority of Angiosperms have distal aperture uniapertunate pollen grains are characteristic of the Monocotyledons and some members of Magnolidae.

For solving some taxonomic problems at lower level the apertural characters are employed significantly. In the genus *Sesamum,* the number of colpi is specific to species, i.e, eleven *S. indicum* nine in *S.prostratum* and eight in *S.laciniatum.*

The two species i.e. *Butomopsis umbellatus* can be differentiated from *B.lanceolata* by the nature of the apertures. *B.umbellatus* is having monocolpate and *B.lanceolata* having penta poratae. Nair (1974) differentiated *Caltha palustris* var *normalis* and var *alba* based on their aperture. (a) var *normalis* is represented by tri-colpate grains and (v) var *alba* by Pentaporate ones.

Exine Sculpturing

The exposed outer surface of the pollen constitute the sculpturing. Some of the important types are *Psilate* (Smooth) *foveolate* (pitted) fossulate (groved) scabrate (very fine projections) *verrucate* (Warty) *baculate* (rod like elements) *pilate* (rod like elements with swollen tips) gemmate (sensile piller) echinate (spiny) rugate (elongate elements irregularly (minute perforations) and reticulate (elements forming open net works).

Occassionally the exine patterns have been found useful in differentiating species of a genus. In the genus *Bauhinia, B.racemosa* has reticulate tuberculate whereas *B.malabarica* has spinulate while *B.acuminate* has *Pilate* and in *B.kigoia* striate lastly in *B.retusa* has verrucate pollen grains are noticed (Nair, 1974).

In the tribe Viciae of the Fabaceae the genus *Cicer* is placed has many characters which suggests that this position is anamalous. Clarke and Kupicha (1976) studied the pollen grains of *Cicer*, of the *Vicia* and the allied tribes, Trifoliae and Onoideae. They found that *Cicer* has pollen that is more similar to that of Ononideae than that of *Viceae.* They suggested the transfer of the genus *Cicer* from the viceae to the Ononideae. The tribe Bombaceae of the family Malvaceae has been recognised as a separate family Bombaceae. The palynological studies also supports this treatment. The exine in the Bombaceae is reticulate, whereas it is spinose in most of the Malvaceae.

Pollen Associations

In most of the Angiosperms, the pollen grains occur singly. In some families of angiosperms families, pollen grains are clustered in Tetrads. Such associations are seen in Dicotyledons in diverse families like Winteraceae, Moniniaceae, Anonaceae, Droseraceae and Mimosaceae. In monocots, they are restricted to advanced families like Juncaceae. Extreme cohesion of pollen is not within Asclepiadaceae or Orchidaceae where they form pollinia. The family Berberidaceae is variously circumbscribed by taxonomist. The recent treatments, *podophyllum* has been removed to a separate family podophyllaceae. The pollen grains united in *podophyllum*, whereas they are free in the other Barberidaceae.

Vij and Kashyap (1978) analysed the taxonomic importance of pollen associated in the multipalynous family orchidaceae of the 50 species they investigated, single pollen grains were found in the relatively primitive groups (*Pahiopedillum villosum Cyperpeduin cordigerun, Cephalanthera ensifolia*) tetrads in the tribe Neottieae and Epidendrieae and Perfect massulae in Orchidae and some Neottieae are recorded Poddubnaja Arnoldi (1976) indicated that pollen grains are single in *Cypripedium* insigne, tetrad in *Calanthe veitovchii* and massulae in *Platanthera bifoliata*. Pollen grains are single in dendrous form and in some monandrous forms *Cephalanthera vanilla*. They remain attached in tetrads as in *Spiranthes australis*.

A. achalensis, Bromhedia finalysoniana, Calantha and *Polystachya flavescens* or form massulae as in *Epigonium roseum, Liparis Viridiflora, Platanthera* and *Peristylus spiralis* or pollinia as in *Monondrae* are noticed (John et al, 1992).

In the family mimosaeae the pollen grain are shed singly, in tetrads, or in pollinia. In *Neptunia, Leucenae prosopis* and

Desmanthus, pollen grains are shed singly. In *Mimosa pudica* the pollen grains are shed in tetrads but in *Mimosa hamata* pollen grains are shed in single tetrad as well as twin tetrads or pollinia.

They are shed together in masses of eight or sixteen pollen grain A in Acacia dealbata, *A.farnesiana* and *A.leucophloea*. All the microspores (in sporongium) together form a single pollinium in *Albizzia lebbeck, Acacia baileyana* and *Caliandra*.

Systematic value of Pollen Nuclear Numbers

The systematic value of the nuclear condition, of the pollen grains is considered by Dufou (1961) At the time of anthesis Angiosperm pollen is 2 nucleate or 3 nucleate . In a survey it was found that monocotyledons (Liliaceae) studied were 2-nucleate, the polypetalous Dicotyledons 2-nucleate, gamopetalous - Dicotyledons 3-nucleate and apetalous members often 2-nucelate. There were however exceptions to these the *Centrospermae* for example were uniformly 3-nucleate. The 3-nucleate condition is regarded as more highly evolved than the 2-nucleate. Although the presence of three or two nuclei in the pollen was generally constant within a family, some families were heterogenous. It would be premature to draw any conclusion.

However Brewbaker (1967) conducted extensive analysis on 2000 species belonging to 265 families. According to Brewbaker two nucleate pollen grains are more primitive in Angiosperms as a whole and within individual taxa three nucleate pollen grains have arisen from the two nucleate ones.

In Monocotyledons, the 3-nucleate pollen grains occur in highly reduced wind pollinated groups; aquatics and a chlorophyllous saprophytes. However in Dicotyledons there is no definite pattern. Certain orders e.g. caryophyllales Plumbaginales, Polygonales and Asterales, have consistently 3-nucleate pollen grains (Dahlgreen, 1975) Magnoliales, Laurales and Violales have consistently two nucleate pollen grains. Most other orders are quite heterogeneous in having both two and three nucleate ones. Both these types occur in one and the same genus as has been observed in *Burmannia,* Lobelia and Ipomoea (Brewbaker, 1967).

Brewbaker (1957) made a significant discovery of connection between bi and trinucleate pollen grains correlated with types of self in compatibility systems opens up further possibilities. In Dicotyledonous taxa particularly the groups with binucleate pollen grain tends to have the gametophytic types of self incompatibility and those with trinucleate pollen grain the sporophytic type. Kooinaan (1972) in Lamiaceae has also demonstrated the correlation between binucleate and tricolpate pollen grains and between trinucleate and hexacolpate pollens grains.

Cytology in Relation to Taxonomy

A special role is claimed for cytological data in taxonomy since the chromosomes are the seat of the heriditory material. The cytotaxonomical data are also denoted as Karyomorphological of chromosomal data on account of usage of chromosomes; then any other inclusion of cell or nucleus. The cytological data are used in two ways (a) Number of chromosomes and (b) their morphological characters are taken as cytological parameters and chromosome behaviour. From an interpretative view point this claim is justified, but chromosome number and morphology which form the common and basic phase of cytotaxonomy. The importance of *chromosome number* as a taxonomic character is that it is probably one of the most constant single feature employed, all individuals within a species usually have the same chromosome number, although there are exceptions. The cytological data are used in two ways (a) Number of chromosomes and (b) their morphological characters are taken as cytological parameters and chromosomes behaviours at meiosis.

Chromosome numbers recorded for spermatophytes vary from n=2 in *Haplopappus gracilis* (compositae) to n=154 in *Morus nigra* (Moraceae) probably around n=250 in *Kalanchoe* species n=263-5 and in *Poa littorosa* (Hair and Beuzenberg, 1961), *Kalanchoe* and *Echeveria* (both Crasulaceae) species show 500-520 chromosome and in *Ophioglossum reticulatum* 1260 chromosomes are there in somatic cells.

In higher plants there are three main different cytological types Diploids, Polyploids and Aneuploids. In *Pinus* and *Quercus* have n=12 chromosomes in all the known species and can be delimited from other related genera on the basis of chromosome number. Throughout the whole group no deviation in chromosome number is found.

The basic number of chromosomes (x) is found to be exactly multiplied in various members of a group giving polyploid series. The common lowest diploid number (2x) by doubling may produce tetraploids (4x) and Octoploids (8x) Triploids (3x) and hexaploids (6x) when somatic chromosome number is an exact multiple of the basic number are termed euploid. An example of polyploid series is found in *Taraxacum* with 2n=16, 24, 32, 40, 48. Chromosome numbers have frequently proved useful within the family at tribal and generic levels. The case of Ranunculaceae, mostly with x=8 but some genera with x=7 and segregated into a separate tribe. The chromosome number vary from 14 to 28, 42, 56, 84 and also from 16 to 32 in *Chlorophyton* that belongs to Liliaceae etc. Nevertheless from the detailed analysis of the meiotic behaviour of chromosomes in one of the species *C.laxicum*. Naik (1976) deduced that the base numbers 7 and 8 can be considered as secondarily most likely derived from primary basic number x=4.

The basic number is defined as the primitive or original number from which polyploid numbers have been derived (Samson, 1958). In some cases the diploids may no longer exist or recognised as such and only apparently polyploidy numbers are reported within the genus concerned. There may be internal evidence of polyploidy which helps to confirm inferences or there may be no evidence as in *Zea mays* (n=10) where the haploid number of 5 in the related genera coix and *Sorghum* suggest strongly that 5 is the ancestral basic number (Darlington, 1956).

The concept of basic number of chromosomes at the level of genus and family has played a prominent role in shaping the prevailing concept of evolution. In practice, the basic number of genus is calculated from the lowest diploid number known in the groups (Love and Love, 1961). In some genera diploid do not exist at all and only polyploids are known to exist. In such cases a theoretical base number has to be inferred (Davis & Heywood, 1963).

The basic number has been considered to be a very dependable and stable marker of direction of evolution. Diminition of the base number is in the indication of progress of evolution. Stebbins (1950) studies in *Crepis, Haplopappus* and many other genera have agreed with this contention. It would appear that reduction in number is very much more frequent than increase, but has emphasized the reversibility of the process. Polyploidy raise of total chromosome of number to very high levels and further progression is frequently by reduction (Jones, 1970). The loss of whole chromosome segments without serious consequences other than a possible upsetting of the generic balance might be expected in the subsequent evolution of the amphidiploid. Examples of this drop in chromosome number are cited by Darlington (1956) who termed this phenomenon as Polyploid drop.

Raven (1975) proposed chromosome numbers are only of limited utility for tracing the evolutionary relationship and for taxonomic placement of the categories. The basic number x=7 is common throughout dicotyledons, but x=11 to 14 is so frequent that this cannot be used in assessing relationship. This can be used only to confirm or to negate a particular suggestion of affinity. Very unusual basic number would not be of any great taxonomic help. Fouquieriaceae with x = 12 has been variously classified by various authors. Hutchinson (1959) and Cronquist (1968) placed it in Tamanicales and Thorne (1968) in Solanales. Both Tamaricales and Solanales which are otherwise thought to be unrelated have the same x = 12 in common with Fouquieriaceae. But there are instances where base numbers have been of good help. Melchior (1964) groups Myricales x = 16 and Juglandales x = 16 into a single order.

There are various mechanisms by which increases or decreases in chromosome number may be brought about, such irregularities occurred during division of the cell or changes in the chromosomes themselves (fragmentation, misdivision of the centromere).

A frequent method of chromosome decrease is brought in through unequal interchange between chromosomes. This has been documented in *Crepis*, where the main trend in chromosome number is a progressive decrease in basic number from 6 to 3, step by step, one chromosome at one time. A decrease from n=4 in *Crepis neglecta* or some related species to n=3 in *C.fuliginosa* appears to have arisen through an unequal reciprocal translocation, so that one chromosome of *C.neglecta* (or a related ancestor) lost nearly long arm. The chromatin attached to the remaining centromere and the centromere itself is not found in *C.fuliginosa* and has presumably been lost (Togby, 1943; Babcock, 1947). A similar deviation of *C.kotschyama* (n=4) from *C.foetida* (n=5) has been described by Sherman (1946) and other examples, in the *Cichorieae* are given by Stebbins *et al* (1953).

The basic set may on the other hand, remain the same though supernumerary chromosomes are added to it. There are two mechanisms :

Polysomics and Monosomics

Polysomics - is the type of aneuploidy where irregularities at meiosis result in cells which are deficient for a particular chromosome or have it duplicated n-1 or n+1 from these complimentary cells, individuals with 2 n-1 and 2n+1 will arise on fertilisation by nominal genetics. Individuals with 2n+1 are termed trisomics, those with 2n-1 are termed monosomics.

Trisomics are polyploid for a particular chromosome.

Trisomics are known from several species and have been studied in detail in *Nicotiana* (Smith 1943) *Lycopersicon* (Rick &

Barton, 1954) and *Zea* (Rhoades and Mc. Clintock, 1935) as well as in economically less important groups such as *Antirrhinum Rudorf Lauritzen,* 1958. *Datura* (Solanaceae) found to be classical example. Which has n=12 - thereby twelve possible trisomics were observed. Blaskeslee (1922, 1930, 1934) produced all these experimentally and identified who also showed that each had multiple phenotypic characteristics which allowed them to be recognised. A certain amount of imbalance is produced by these extra chromosomes, but the capacity of the plants are not greatly affected. The addition of two extra chromosomes (tetrasomes) does however seriously weaken their viability and evidently their powers of competition in nature. In tetraploids the disequilibrium is less marked. The taxonomic and evolutionary significance of these trisomics is limited by their genetic inbalance. Although phenotypic differences are comparable to those found in separating species. Diploid monosomics are normally inviable and are seldom found. Aneuploids that do occur most likely are polyploid where the loss of chromosome can be more easily assimilated than by a diploid as pointed out Allopolyploidy has been most studied group of economic importance especially in cultivated plants. The latter include several allopolyploids that have arisen from crosses between different genera. *Triticale* 2n=$\pm$ 28) from *Tritivum* aestivum (n=21) x Cerale secale (n=7); *Raphano brassica* (n=18) from *Raphano sativum x Brassica oleraceae* (n=9) Clasusen *et al.* The most complex situation of this type is observed in the cultivated bread wheats - *Triticum aestivum). Triticum aestivum* is an allohexaploid groups found to contain genomes (AA BB DD) derived through chance hybridisation between two genera. *Triticum* and *Aegilopsis* this probably involved crossing between the tetraploid (AA BB) wheats and this diploid DD genome species *Aegilopes squarrosa.*

Differences in chromosome number within a species complex are sometimes, but certainly not always correlated with differences in morphology (Klippius, 1967) *Galium mollugo* and *G.veruma* have two different cytotypes - diploids and tetraploids, but are morphologically indistinguishable (Davis and Heywood, 1963) Mohinder Pal (1962) reported two cytodemes in *Pupalia lappaceae* (Amaranthaceae) diploid and tetraploid.

There are several taxa which exhibit disploid variation in their chromosome number. *Listera orata* has got 2n=34, 36 and 38 while *reveneana* has 2n=35, 40, 45, 46 and 45. The aneuploid complex in Iseilima is another example. *Iseilima anthephoroides* n=5, I.*venkateswarli* n=4 and *I.lubbardi* n=3 (Satyavathi et al, 5,6,7 and 8 Hilborne, 1931) but the chromosomal differences are not accompanied by breaks in morphology.

Karyotypic Information

Karyotypic data especially from cytological studies have been shown to be useful in taxonomy at all levels upto that family and even above, but at species level and groups of species. The most obvious characteristics of Karyotypes are the size and shape of chromosomes relative to each other within the same compliment. These differences are often represented in a diagramatic form known as an idiogram or karyograms.

Viewed at mitotic metaphase, the individual chromosomes of some groups of plants show marked differences in size and shape. In other cases the chromosomes may be quite uniform differences may be gradual between the various chromosomes. The constant position of the centromere also allows chromosomes to be characterized by the relative length of the arms. Further differences may be shown by the presence and position of secondary constrictions marking off bodies known as satellites when they are distal. The most familiarly utilised aspect of chromosome structure is position of centromere i.e. arm length ratios of each chromosome in genome. When the centromere is middle of the chromosome are known as metacentric whereas those with centromere near one end of the chromosome are known as acrocentric, while the ones with a truly terminal centromere are known as telocentric. Conspicuous differences in appearances of Karyotype have been found in species having same chromosome number in many genera and families of flowering plants. These differences follow definite trends associated with trends of morphological specialization in some instances. For example, the chromosomes in the Karyotype differ little from each other in size and most of them are V shaped with median or submedian centromere which were observed in the primitive genus *Helloborus* (Ranunculaceae). The most common ones are karyotypes of this nature, Lavitsky (1931) proposed them as generalised types. Various specialised types have been proposed to have been derived from these. Thus the chromosomes with uniform length may gradually differ in size and shape from V-shaped (metacentric) to 'J' shaped with long and short arm and ultimately to rod shaped or I shaped (Telocentric) and become asymmetrical. In the advanced genera *Aconitium* and *Delphenium* (Ranunculaceae) for example the flowers are zygomorphic with the largest number J-shaped chromosomes. The Heterochromatin chromosome segments or chromosomes which remain contracted and deeply staining or contract early in contrast to the behaviour of the euchromation in the resting and prophase stages. Heterochromatin may also despiralize and appear as lightly stained segments at metaphase when euchromation is not condensed. The genetic content or at least the function of the heterochromatic regions of chromosomes have been shown to be quite different from the euchromation and there appear to be a number of different kinds of heterochromation. Heterochromatin chromosomes and segments are useful in characterizing the morphology of particular chromosomes and karyotypes.

Jones (1970) demonstrated in certain membes of the Commelianaceae the reverse trends of specialization i.e. from telocentric chromosomes to metacentric chromosomes were less common.

At all levels of the taxonomic hierarchy chromosome structure, together with chromosome size, and number have been found extremely useful, Jackson (1971). The best known use of the importance of chromosome size is known in the Yucca monocotyledons genera Yucca, Agave and relatives. *Yucca* and some others have flowers with superior ovary while *Agave* and others have an inferior ovary. *Yucca* was placed in Liliaceae and *Agave* was placed in the Amaryllidaceae with the above reasons. The great overall similarity prevails in these genera based on this now they are usually segregated from their respective families and placed in the *Agavaceae*. When it was found that Agavaceae are characterized by a very distinctive bimodal karyotype consisting of five large chromosomes and 25 small ones, the evidence is used to substantiate the classification. (McKelvey and Sax, 1933). The finding also accorded by Hutchinson is realignment of these families and his separation of both these genera into a distinct family Agavaceae.

The use of chromosome number and morphology has solved many taxonomic problems at the generic sectional and specific levels and show how many species are inter-related in the genus *Crepis* (Babcock, 1947). The chromosome number is constant within a genus, this is useful in generic limitation. In the genus *Tephrosia* (Fabaceae) all species are with 2n=22 - except for *T.constiecta* with 2n=16. This species has been separated as a distinct genus *Sphinctrospermum* and chromosome number supports this separation.

Meiotic behaviour of chromosomes in Taxonomy

Valuable information can be procured from a study of the way the chromosomes behave at meiosis. It may permit a distinction to be made between taxa as has been demonstrated by Garber (1958) in the genus *Collinsia* (Scropulariaceae). He employed chiasma frequencies as a means of grouping species within the genus those with sessile flowers possessed a low mean number of chiasma while those with pedicillate flowers had a relatively high number of chiasmata. A classification of organisms according to the behaviour of chiasmata at metaphase is given by Darlington (1937) such behaviour may give some indication of affinity.

More frequent use is made of meiotic behaviour as a means of indicating relationships through the kind and degree of pairing that takes place. Since it shows that hybridization has occurred, indicate chromosome structural differences in the parental chromosomes such as reduplications or explain the cause of sterility, great weight is often placed on the degree of chromosome homology in hybrids as an estimate of the degree to which the parental species are related. The chromosomes derived from two parents are often not homologous at meiosis and do not therefore pair, if the numbers of the parents are different, pairing will of course be incomplete.

To distinguish morphologically between chromosomes, at least in favourable material has been greatly enhanced by new staining techniques using **Giemsa** and **Flurochrome** dyes which stain chromosomes in consistent banding pattern instead of with a uniform intensity as in the case with the usual basic fuchsin (Fuelgen reagent) Scheweizer and Ehrendorfor (1976) Ehrendorfer *et al* (1977) have applied this technique in the systematics of the genus *Anacyclus* and compared and correlated with results obtained by morphological and phytochemical approaches. The data also well related with phylogenetic conclusions derived from other fields of study and suggest a more positive correlation for perennial taxa of *Anacyclus* than for annual one. *A.officinarum* was exhibit to be probably a hybrid between *A.pyrethrum* and another annual species *A.radiatus.*

Chemotaxonomy

Plant chemistry and biochemistry will provide a wide new source of characters. Plant chemistry and biochemistry will provide a new source of taxonomic characters. Employing chemical data for explaining taxonomic problems is variously known as chemosystematics, biochemical systematies and chemotaxonomy. Actually chemotaxonomy makes the taxonomy more meaningful than ever before.

Chemical characters of plant can be widely applicable in

classification as do characters from gross morphology. The chemical characters provided useful information for understanding of evolutionary relationships of various taxa and have also helped in the solution of many taxonomic problems. Various types of chemicals that occur in plants form the taxonomic evidence. But all kinds of chemical substances do not reveal information useful to the taxonomist. All chemical constituents are not having equal taxonomic importance. They can be conveniently be divided into macromolecules and micromolecules. DNA, RNA and proteins also known as primary, secondary and tertiary semantides (based on their metallic activities) are macromolecules. The chemotaxonomic studies involving them, include DNA hybridization, DNA and RNA hybridisation patterns etc. The micro-molecules are numerous and can be divided into primary metabolites and secondary metabolites. The farmer comprises free aminoacids, simple sugars etc. From among free aminoacids, the protein aminoacids are less taxonomically significant, since they are subjected qualitative and quantitative changes due to environmental rigors. They are present in all the tissues and vulnerable for changes in the tissue in their respective tissue. Chemotaxonomic characters like any others, are useful at all levels of the taxonomic hierarchy and it is not possible to predict the level of their value, flower colours are considered as an example. For flower colour often varies within species especially where garden varieties have been selected, red, blue, yellow, white and all combination and intermediates can exist in one species. Occurrence of secondary compounds of low molecular weight such as 400 non protein aminoacids phenolics and betalins, alkaloids, terpenoids and steroids provide valuable clues to systematist, Chemical evidence used by the taxonomist at all levels of the hierarchy of classification from varietal rank to the divisions of plants. Chemical data are also became a reliable guide to phylogenetic relationship of by various taxa. Some of the important chemical sources of taxonomic evidence are discussed briefly as follows.

Aminoacids

Two groups of amino-acids protein amino-acids and non-protein amino acids are present in plant system. The protein amino-acids are building units of proteins and they are released on hydrolysis of proteins. The non-protein amino acids which are not so far found in combination in proteins are more numerous and are estimated to be about 300 in number. Interruption of occurrence and less susceptibility to rapid change increase the taxonomic value of non-protein amino-acids.

Occurrence of non-protein amino-acids in Fabaceae has been attracted the considerable interest to the taxonomist. They are present in very high concentrations in the seeds. Bell (1962) has identified seven infrageneric groups in the genus *Lathyrus*. Each group is represented by a different amino-acids or group of amino acids. In the same way four infrageneric groups have been recognised in the genus *Vicia*.

The amino acid content of seeds in the section Gummiferae of the genus *Acacia* is so different from that in the other sections that the species of this section can be recognised on the basis of this single character alone. Free amino acids lathryine is so far known only in the genus *Lathyrus* and this supports the present delimitation of the genus. Azetidine-2-carboxylic acid is extremely rare in occurrence and is characteristic of many genera of the Liliaceae Amaryllidaceae and Agavaceae which are related families.

All the chemical contituents of a plant are potentially useful to a taxonomist but in reality some sorts of molecules are far more valuable than others. In addition from inorganic compounds, which are of relatively little use, three main varieties of compounds can be identified. Primary metabolites secondary metabolites and semantides.

Primary metabolites are parts of important metabolic pathways and most of them are of universal occurrence or at least present in wide range of plants. Aconitic acid (isolated from *Aconitum*) or critic acid (from *Citrus*) for example participate in the Krebs (tricarboxylic acid) cycle and are present in all aerobic organisms. The occurrence or non occurrence of such compounds is therefore not of much systematic values. It is apparently very clear that 22 or so, amino acids which are known to be constituents of plant protein or any of the sugars which occur in photosynthetic carbon cycle or so on. Although in some cases the quantities of such metabolites vary considerably among the taxa and this in itself can be taxonomically useful universally occurring substances were first detected in taxa, often possess particularly large quantities of the molecules concerned well above the amounts which participate in the essential metabolic pathways after as food storage materials. At times such compounds are stored in a different form from that in which they are metabolized for example sedoheptulose, a sugar constituting the carbohydrate food reserve of the genus *Sedum* which as sedoheptulose diphosphate is part of the photosynthetic carbon cycle.

Secondary metabolites perform nonvital functions and are therefore less widespread in plants. Nevertheless the rare occurrence of these among plants which renders them valuable taxonomic information. The very familiar groups of compounds which have been utilized in this way include alkaloids, phenolics, glucosinolates, amino acids, terpenoids oils, waxes, and carbohydrates.

As a whole the secondary metabolites are largely waste substances, food stores, pigments, poisons, scents, structural units or water repellents etc. On the whole they do have essential function but of a general nature so that precise configuration of the compound is not very important. Thus a yellow pigment with an absorption maximm at 477 nm might be a betalinor anthocyanin and a poison might be an alkaloid or a glycoside.

Semantides are the information carrying molecules DNA is primary semantides, RNA secondary semantide and proteins are tertiary semantides. These three are precisely known as *trinity of life*. They follow the sequential transfer of the genetic code from the primary genetic information (DNA). In reality the sequence of nucleotides and aminoacids in these substances should provide all the taxonomic information necessary for classification and offer an alternative to study the secondary metabolites, cytology, morphology and anatomy etc. Since the later are manifestations of the former. In the gathering of the sequencing data great difficulties are encountered and the results present baffling anomalies. At times the samantides together with the larger polysaccharides are known as macromolecules and the primary and secondary metabolites as micromolecules.

The role of chemical characters in plant taxonomy

The important role played by chemotaxonomy can be seen when chemical characters correlate well with data obtainable from other sources. To cite one example the chemical discontinuity appear among the families of the order Rhoeadales as defined by Wettstein (1935 Table 14.7) has been used in modern taxonomists as a reason for dividing this assemblage of six or seven families (Cronquist, 1968, 1969). For making this split lie in substantiating it with anatomical differences and correlating with morphology. The new categorization is also suggested by recent serological and palynological studies of the relevant families.

It should be remembered in this case, that there are many morphological similarities between the separated groups. For example the common floral features such as perfect hypogynous flowers. Compound ovaries with parietal placentation. The additional chemical evidence may have crucially substantiated in the acceptance of this newer grouping. Relatively wide chemical surveys have been conducted in the relevant families based on good chemical data. Over 300 Cruciferae species have been examined for glucounolates with positive results (Kjeer, 1976) along with all Papavaeraceae studied have been shown to have complex mixture of alkaloids in their tissues (Manske, 1966).

Table 14.4 : A summary of some groupings of natural products which have been used in chemotaxonomic correlations including some comments on the botanical distribution of these constituents (Radford et al., 1976).

Compound Grouping	Comments on botanical distribution	Citation
Terpenoids	Widely distributed	Weissmann (1966, see: swain)
Monoterpenes	Widely distributed, of taxonomic value primarily below the generic level	numerous reports, see Mirov (1967) or others
Sesquiterpenoids	rather wide distribution, but particularly important and useful in the taxonomy of the Compositae	Herz (1968, see: Mabry et al.) Herout and Sorn (1969, see: Harborne and Swain)
Asperulosides and Aucubins (Iridoid glycosides)	Rubiaceae, Scrophulariaceae, and related families, Plantaginaceae, Cornaceae, and others.	Bate-Smith and Swain (1966, see: Swain) Kooiman (1969, 1970)
Ranunculins	found only in the Ranunculaceae	Bate-Smith and Swain (1966, see: Swain) Kooiman (1969, 1970)
Cyanogenetic Compounds	Ranunculaceae and other families	Ruijgrok (1966, see: Swain)
Polyalcohols	widely distributed but have taxonomic potential	Plouvier (1963, see: Swain)
Sulfur Compounds	a chemically diverse grouping of compounds which in their various forms are widely distributed, the isothiocyanate-producing glycoides are characteristic of families in the Rhoeadales	Kjaer (1966, see: Swain)
Amino acids (non-protein)	Liliaceae and related families, Leguminosae (particularly Papilionate, and other groups)	Bell (1966, see: Swain)
Alkaloids	A chemically and biosynthetically diverse group of compounds rarely found in the lower vascular plants and irregularly distributed among the angiosperms; highly useful in the taxonomy of some groups	Hegnauer (1966, see: Swain) Schafer (1964) Price (1963, see: Swain)
Betacyanins And betaxanthins	Centrospermae ("Betanales")	Mabry (1966, see: Swain)
Alkanes (fatty acids And waxes	widely distributed, possibly of use in classification below the genus level, and of use in organic geochemistry	Shorland (1963, see: Swain) Douglas and Eglinton (1966, See: Swain)
Fatty acid epoxides	found in seven families	Bu'lock (1966, see: Swain)
Acetylenes	distributed in the basidio-	Bu'lock (1966, see: Swain)

Table 14.4 Contd..........

Compound	Comments on botanical grouping	Citation distribution
	Mycetes and atleast 13 angiosperm families of particular use in the link between the Compositate and the Umbelliflorae ferns	Srensesn (1968, see : Mabry et al)
Assorted compounds		Berti and Bottari (1968, sce : Reinhold and Liwschitz)
Diterpenses	Rather wide distribution in the seed plants, taxonomically useful in specific groups such as the conifers	Edrtman (1968, see : Mabry *et al*) Ponsinet et al (1968, See : Mabry *et al*)
Triterpenses	Wide distribution in living organisms, taxonomically useful in several angiosperm families such as the Cucurbitaceae, and in the fungi.	Posinet et al (1968, see : Mabry *et al*)
Carotenoids	The universally distributed photosynthetic carotenoids useful in algal classification; the non photosynthetic carotenoids in fruits of possible limited value as taxonomic markers in the angiosperms.	Goodwin (1966 see Swain)
Flavonoids	Very large number of diverse compounds found throughout the vascular plants and in nearly, if not all angiosperms; of great taxonomic value below the genus level and of possible use in the classification of higher categories	Bate-Smith (1963; see : Swain); Harborne (1966, See : Swain); Wagner (1966, see : Swain); Williams (1966, see : Swain); Alston (1968, see : Mabry et al) extensive literature
Lignins and Liganas	Useful in the classification of higher categories	Endtman 1968, see; Mabry *et al*
Quinones	Widely distributed among living organisms, but compounds of this type with limited distribution have potential value in the classification of a number of angiosperm families	Mathis (1966, see : Swain)
Polysaccharides	Universally distributed, probably useful in the classification of higher categories particularly among the algae comparative data mostly lacking	Percival (1966; see : Swain)
Plant glycosides	Great chemical variation in the non-sugar portion of the molecule, varyingly of great usefulness in taxonomy	Paris (1963, see : Swain)
Assorted compounds	Lichens	Huneck (1968, see : Reinhold and Liwschitz)
Vide range of chemical approaches	Umbellieferae	Heywood (1971)
Vide range of chemical approaches	Leguminosae	Harbone et al (1971)

The phenolics molecules frequently studied in the chemotaxonomy fall into a general class known as flavonoids. These are further classified into flavonones, isoflavonoids, flavonols, anthocyanins etc. One of the secondary metabolites best projected are flower pigments. The flower pigments which are usually anthrocyanins and anthoxanthins which are always under genetic control. But at the same time all variation in pigment production is not genetic origin.

Bate Smith (1948, 1958, 1962) succcintly emphasized the early application of evidence from phenolics to taxonomic problems. According to him leucoxanthins are generally present in woody plants. The herbaceous Primulaceae, however are exceptionally rich in leucoanthocyanins and some of woody members in Oleaceae and Scropulariaceae lack them. By using the phenolic characters Bate Smith (1982) utilized these for rearrangement of the species *Iris*. William (1982) utilised the chemical evidence from flavonoides in the classification of *Malus*. On the basis of flavonoids pattern Meclure and Alston (1966) identified highly problematic taxa of Lemnaceae in which morphological and cytological characters were observed to be inadequate in differentiating different taxa of *Spirodela, Lemna, Wolfia* and *Wolfiella* William (1976) compared leaf flavonoids in Liliaceae, Juncaceae. Cyperaceae and Poaceae and supported the view that all these four families have arisen four Liliaceous ancestros. In table 14.5 some major groupings of plant natural products are listed alongwith some comments on their botanical distribution.

Table 14.5 : Revised Classification of the families of the order Rhoedales (Sensu Wettstein)

Families	Chemical Characters	Biological characters
CAPPARALES Brassicaceae Capparaceae	Glucosinolates (mustard oil glycosides), Universal; isoquinoline alkaloids absent	Centrifugal stamens; no latex system
Resedaceae Moringaceae		
PAPAVERALES Papaveraceae Funmariaceae	Glucosinolates absent : Isoquinoline universal	Centripetal stamens; well developed latex system

Williams *et al* (1981) opined Araceae to be unusual amongst Monocots because of their simple and relatively uniform flavonoids profile. Such sub families were not clearly distinguishable and it was only at the tribal level where some significant taxonomic patterns were observed. From observations they point out that this has not been derived from Liliaceous stock since flavone C-glycosides, the characteristic constituent of the *Aroids* rarely encountered in the rpesent day Liliaceous taxa. Further they were observed to be totally absent from the tribe Aspidistreae from which the origin of this family has been suggested by Hutchinson (1959).

Chemistry is not only useful at the generic level but also in family assignments. There are still considerable number of usually small angiosperms families whose affinities are not clear and which are difficult to place with certainity into an ordered system of classification. The taxonomic relationship of Julianaceae family became subject of controversy for at least a century. Based on wood anatomy and morphological features associated with wind pollination, the family has variously been placed in three different associations near either Anacardiaceae or theBurseraceae of the Juglandaeae.

Young (1976) made chemical analysis of heartwood of *Amphipteryguim adstrigens* (Julinaceae) yielded sixteen flavonoids, seven of which were 5-deoxy flavonoids. These are characteristic substances occur in Anacardiaceae and one of them the aurone *rengasin* is known otherwise from this family. Conversely no evidence could be found in Amphipterygium Juglone which is characteristic quinone of *Juglandaceae*.

Analysis of flavonoids carotenoid and quanonoid pigments in plants of the Gesneriaceae have shown significant correlations with the fitting in with subfamily division of Burtt based on geography and anisocotyly, while earlier division of von Fritch based on the position of ovary (Harborne, 1967). Similarly phytochemical surveys in the tribe Anthemidae (Asteraceae) for both flavonoids and polyacetylenes have indicated that the morphological splitting of the *Chrysanthemum* complex Sensu lata into smaller generic groups is justified on chemical grounds (Bohlman et al, 1964; Harborne, et al 1970) considering the taxonomic instances of compounds with unique distribution in plants, the study of the "nitrogenous anthocyanins" is perhaps the best known example. The Betalins as they are known, include the red to violet betacyanisns and the yellow betaxanthins (Fig 14.2)

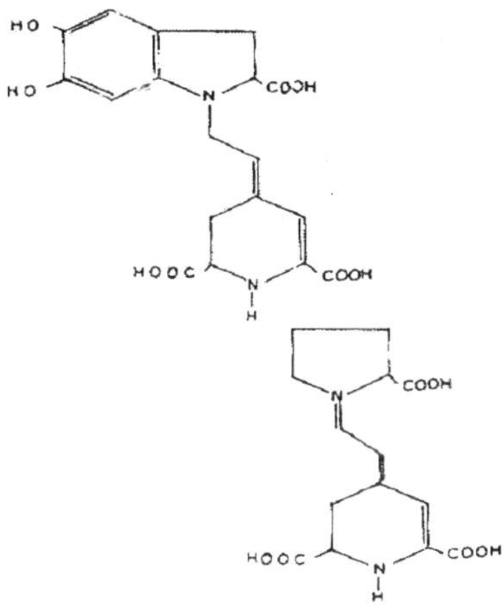

Fig. 14.2. Betacyanin and Betaxanthin

Betalins have been found in nine families of Angiosperms, all members of the order centrospermae. These are Aizoaceae, Amaranthaceae, Basellaceae, Cactaceae, Chenopodiaceae, Didieraceae, Nyctaginaceae, Phytolaceae and Portulaceae (Table 14.6).

The term cynogenesis defined as the ability of certain plants to release hydrocyanic acid after injury of cells. Cynophoric plants usually contain one or several cyanogenric glycosides when Bohm (1801) first detected HCN and amgdalin (the first plant glycoside) in seeds *prunus amygdalus* (Rosaceae) many Rosaceous and other plants have been shown to be cynophoric. About 2056 species of plants are known to be cyanophoric till date. Cynogeneric taxa are frequently occur in *Araceae, Junceceae, Juncaginaceae Graminaceae* and *Scheuchizeriaceae* among Monocotyledons and in several

Dicotyledonous families of Angiosperms. According Hegnauer, (1977) cynogenesis if adequately and carefully used as a character in plant systematics may prove to be of significant value in the classification of plants even at the higher systematic value. In the *tracheophytes* the distribution of individual cynogenic compounds supports the theory that *Liliospida* (Monocotyledons) from ancestors resembling present day magnoliidae since cynogenesis proceeds in exactly the same way in both these groups. With only triosine as the precursor Hegnauer, 1973, Van valen, 1978). More than one biosynthetic group of cyanophoric compounds occur only in very large genera and families belonging to Dilleniidae, Rosidae and Asteridae based on the facts known today. The accumulation of cynogeneric compounds in these taxa is taxonomically significant at the infra-familial level. Linamarin for examples is a character of many loteae, Trifolieae and Phaseolae in Fabaceae and in Asteraceae.

Table 14.6 : Classification of the order centrospermae : (Caryophyllales) Varying according to the use of evidence from pigments.

Structural classification Engler & Prantl (1897 – 1915)	Chemical Classification		Compromise classification
Centrospermae	Cheopodiales	Order :	Caryophyllales
Amaranthaceae	Aizoaceae	Suborder :	Chenopodinae
Aizoaceae	Amaranthaceae		Aizoaceae
Basellaceae	Basellaceae		Amaranthaceae
Caryophyllaceae	Cactaceae		Basellaceae
Chenopodiaceae	Chenopodiaceae		Cactaceae
Didiereaceae	Didiereaceae		Chenopodiaceae
Molluginaceae	Nyctaginaceae		Didieraceae
Nyctaginaceae	Phytolaccaceae		Nyctaginaceae
Phytolaccaceae	Portulacaceae		Phytolaccaceae
Portulacaceae			Portulaccaceae
	Caryophyllales	Suborder	Carophyllinae
	Caryophyllaceae		Caryophyllaceae
	Molluginaceae		Molluginaceae
Cactales			
Cactaceae			

Several botanist of India have contributed to the chemotaxonomy of plants. To mention a few among them are S.D. Sabinis from Baroda I.L. Narayana from Warangal and M.Radhakrishnaiah, from Hyderabad. According to Nageswar *et al*, (1984) the distribution of a different chemical compounds in *Caesalpinia* and form for key or identification of the species on chemical grounds.

On phytochemical grounds Radhakrishnaiah et al, (1982) concorded the separation of Trapa from Onagraceae. Radhakrishnaiah *et al* (1984) based on the phytochemical studies support to the close relation among the genera. *Sparganum, Typha* and *Pandanus* and their retention under on Englerion taxon *Pandanales* and negate their segregation. Anuradha *et al* (1988) made a study of chemotaxonomy of *Capparaceae* however their study did not support the separation of Cleomaceae from Capparidae, Rajendra Prasad *et al* (1986) felt that the chemical data obtained do not suggest the erection of *Mimosacea* as an independent family.

Isoflavone iridin is known only in pogoniris section of *Iris*. *Iris flavissima* which was originally placed in section Pogon *Iris* does not contain Iridin and other hand resembles species of regelia section in phenolic characters. This favours transfer of *Iris flavissima* to regelia section.

1. Juglones present - *C.sappan*
1. Juglones absent :
2. Aucubin compounds present - *C.coriaria*
2. Aucubin compounds absent
3. *Catechol tannins present :*
3. *Catechol tannis absent* : :
4. Methylene dioxy compounds present - *C.pulcherrima* (*red*)
4. Methylene dioxy compounds absent - *C.pulcherrima (yellow)*
5. *Leucoanthocyanins Present* :
5. *Leucoanthocyanins absent* :
6. Antharaquinones present - *C.gledeschiodes*
6. Anthraquinones absent
7. Tannins present - *C.cacalace*
7. Tannins absent - *C.sepiara*
8. Syringaldehyde doubtful - *C.bonducella*

8. Syringaldehyde absent
9. Steroids present - *C.digyna*
9. Steroids absent -
10. Tannins present - *C.ferrae*
10. Tannis absent - *C.tortuosa*

Table 14.7 : Revised Classification of the families of the order Rhoedales *(Sensu Wettstein)*

Families	***Chemical Characters***	***Biological characters***
CAPPARALES	Glucosinolates (mustard oil gycosides)	Centrifugal stamens; no latex system
Capparaceae	Universal; isoquinoline alkaloids absent	Resedaceae moringaceae
PAPAVERALES Papaveraceae funmariaceae	Glucosinolates absent : Isoquinoline universal	Centripetal stamens; well developed later system

Table : 14.8 : Classification of the Order Centrospermae : (Caryophyllales) varying according to the use of evidence from pigments

Structural classification Egler & Prantl (1897 – 1915)	***Chemical Classification***		***Compromise classification***
Centrospermae Aizoaceae	Cheopodiales	Order :	Caryophyllales
Amaranthaceae Basellaceae.	Aizoaceae	Suborder :	Chenopodinae
Caryophyllaceae	Amaranthaceae		Aizoaceae
Chenopodiaceae	Basellaceae		Amaranthaceae
Didiereaceae	Cactaceae		Basellaceae
Molluginaceae	Chenopodiaceae		Cactaceae
Nyctaginaceae	Didiereaceae		Chenopodiaceae
Phytolaccaceae	Nyctaginaceae		Didieraceae
Portulacaceae	Phytolaccaceae		Nyctaginaceae
	Portulacaceae		Phytolaccaceae
			Portulaccaceae
	Caryophyllales	Suborder	Carophyllinae
	Caryophyllaceae		Caryophyllaceae
	Molluginaceae		Molluginaceae

Cactales

Cactaceae

Table 14.9

Compound	***Comments on botanical grouping***	***Citation distribution***
	Mycetes and atleast 13 angiosperm families of particular use in the link between the Compositate and the Umbelliflorae ferms	Srensesn (1968, see : Mabry et al)
Assorted compounds		Berti and Bottari (1968, see : Reinhold and Liwschitz)
Diterpenses	Rather wide distribution in the seed plants, taxonomically useful in specific groups such as the conifers	Edrtman (1968, see : Mabry et al) Ponsinet et al (1968, See : Mabry *et al*)
Triterpenses	Wide distribution in living organisms, taxonomically useful in several angiosperm families such as the Cucurbitaceae, and in the fungi.	Posinet et al (1968, see : Mabry *et al*)
Carotenoids	The universally distributed photosynthetic carotenoids useful in algal classification; the non photosynthetic carotenoids in fruits of possible limited value as taxonomic markers in the angiosperms.	Goodwin (1966 see swain)
Flavonoids	Very large number of diverse compounds found throughout the vascular plants and in nearly, if not all angiosperms; of great axonomic value below the genus level and of possible use in the classification of higher categories.	Bate-Smith (1963; see : Swain); Harborne (1966, see : Swain); Wagner (1966, see : Swain); Williams (1966, see : Swain; Alston (1968, see : Mabry et al) extensive literature
Lignins and Liganas	Useful in the classification of higher categories	
Quinones	Widely distributed among living organisms, but compounds of this type with limited distribution have potential value in the classification of a number of angiosperm families	Mathis (1966, see : Swain)
Polysaccharides	Universally distributed, probably useful in the classification of higher categories particularly among the algae comparative data mostly lacking	Percival (1966; see : Swain)

Plant glycosides	Great chemical variation in the non-sugar poıtion of the molecule, varyingly of great usefulness in taxonomy	Paris (1963, see : Swain)
Assorted compounds	Lichens	Huneck (1968, see : Reinhold and Liwschitz)
Vide range of chemical approaches	Umbellieferae	Heywood (1971)
Vide range of chemical approaches	Leguminosae	Harbone et al (1971)

Examples from semantides

Serology and Taxonomy : Immune serum (antibody) produces antibodies as a result of introduction of foreign cells or particles (antigenus). Serology is the study of the origin and properties of antiserum. Boyden (1964) defined serology as that portion of Biology which is concerned with the nature and interactions of antigenic material and antibodies. The science dealing with immuno-chemical reactions between serum antibodies and antigens (Kloz, 1971) has proved valuable taxonomic information because its techniques help to detect homologous proteins. Boyden (1967) and Fair brothers (1968, 1977) have thrown some light on the importance of this type of study in plant systematics. Storage proteins are most amenable for taxonomic studies followed by pollen proteins. They are injected in emulsions form into test animals such as rabbit. The nature and amount of the precipitate formed by antigen antibody complex is of potential taxonomic value. A specific antibody is produced in response to specific antigen. The serum is then made to react *in vitro* with the antigenic protein as well as those of other taxa of which the affinities are in question. The amount of precipitation indicates the degree of protein homology and hence it is taken as a phylogenetic marker and taxonomic characters. Kowarski (1901) Bertarelli (1902) and Magnus (1968) were well known serologists. From various grass and legume species they compared proteins showing similarities and differences. Using crude plant protein extracts by serological studies have been widely employed in estimating phylogenetic relationships and illustrating the taxonomy of a wide variety of taxa. Thus a serological comparison of major seed proteins from Angiosperms has pointed out a close proximity among the Magnolidae, Hamamelididae and cornflorae (Sensu, Dahlgreen, 1980) and has refuted the idea of their independent evolution. It has also accorded the homogeniety of the Iridioid producing *Cornflorae* and supported the inclusion of Gentianaceae in it. (Jensen & Grevera, 1984).

Serological studies exhibited that the genus *Liriodendrons* has been found to be quite distinct from other members of the family Magnoliaceae and the genera *Magnolia* and *Michelia* portrayed closest relation within the family (Johnson and Fair brothers, 1965). Identical studies have also been useful in estimating the relationship of the Nymphaceae and Nelumbonaceae (Simon, 1971) generic relationship in the Caprifoliaceae. Hill brand and Fair brothers, 1969, 1970) and in the Rubiaceae and related groups (Lee & Fair brothers, 1978). According to pickering & Fair brothers 1970) the findings support the classification of the family *Umbelliferae* into Hydrocotyloideae, Baniculoideae and Apioidea and suggest that the Apioideae is more closely related to Saniculoideae than Hydrocotyloideae.

Chrispeels and Gartner (1978) supported based on the serological evidence for assigning *Phaseolus aureus* and *P.mungo* to the genus *Vigna*. The serological relationship strongly supported the classification of Ranunculaceae at tribal and generic levels based on classical evidence particularly from the chromosomes (Jenson 1968, 1974). Close relationship of *Delphinnium* and *Aconitum* is supported by serological evidence. *Hydrostasis* which is sometimes classified with Berberidaceae has more close serological similarity to the *Ramunculaceae* than *Berberidaceae Trollis* which is regarded as a link between follicular and achenial fruit first bearing genera in the family, resembles closely with Adoinis. Serological methods have also proved useful in the taxonomy of legumes and potato and in the classification of *Bromus*. Smith (1972) made a study of the relationship of annual brome grasses

(Bromus) relating immuno electrophoretic data to those from morphology, cytology, and cytogenetics. He was able to focus some light on species problems, diploids tetraploids relationships and hybrids and the serological and cytological distinctness of previously named variety of *Bromus secalinus* led him to recognise it as a new species *B.pseudosecalinus*.

DNA and RNA Hybridisation

DNA amount have been a valuable source of phylogenetic and consequently also of taxonomic evidence. But it has so far not proved possible to sequence DNA the larger RNA molecules in routine way and for this reasons nucleic acid characteristics are inferred by indirect means such as DNA hybridisation. In brief DNA extracted from one organism is treated to convert it to a single standard polynucleotide chain and the amount of re-aassociation (annealing) with similarly treated DNA from another taxon which occurs on mixing the two is taken as a measure of similarity of the nucleotide sequences.

The utility and importance of DNA values in taxonomy may perhaps best be demonstrated in *Scilla bifolia* alliance a group of plants with a constant basic number of x=9 and little evidence for major structural rearrangements (Greilhuber and Speta 1977), (Grielhuber 1979) most remarkable from the stand point of traditional systematic is the incorporation of *Chionodoxa* (Boiss). Speta (1976). The second grouping is mostly based on seed colour, which is yellow, brown or black. C-values ranges from 4.2 to 9.4 pg (1c) Higher C-values are correlated with yellow seeds, intermediate values with grey or brown seeds, and intermediate to low values with black seeds. There is no parametric overlap between the taxonomic groups when euchromation estimates rather than total DNA amounts are considered. The group as a whole appears to be monophyletic due to such supposed apomorphies as predominantly bifoliate condition, a particular claisome type a nuclear endosperm. A hypothesis of phylogenetic relationships is documented. In quantitative species comparisons based on DNA hybridisation to and for *secale, Hordeum* and *Triticum* showed that these cereals. Wheat (75%) is more closely related to rye (100%) than in barley.

Ribosomal DNA and RNA sequences are now available for a range of organisms (Brimacombe 1981) RNA has been investigated by the equivalent technique of DNA-RNA hybridization whereby the amount of association occurring between DNA of an organism and a fraction of the RNA (ribosomal RNA) of another is taken as a measure of relationship. Mabry (1976) examined plants of the Centrospermae (Caryophyllates) by this methods and concluded that Caryophyllaceae (contain anthocyanin instead betalins) are quite close to the betalin containing families, but not as close as the latter are to each other.

Amino Acid Sequencing

Amino acid sequencing is a recent trends in development aiming to identify pure proteins down to the atomic level. Now it is possible to break off the amino acids from polypeptide chain one by one identify each in turn chromatographically and so build up the complete sequence of amino acids step by step. A first step is to break the total polypeptide chains into smaller peptide fragements each of which is then sequenced separately. Previously these small peptide units were separated by chromatography and / or electrophoresis (finger printing) and such relatively crude results themselves proved of taxonomic value. The variation in the precise sequence of amino-acids in a single homologoous protein ie one presumably of monophyletic origin through out a range of organisms. This relies upon the fact that a particular protein does not have or single invariable structure; but a good proportion of it may vary without altering its essential functions. If one cites cytochrome 'C' for example (the molecule most used for sequencing purposes) about 79 out of approximately 113 amino acids vary from species to species but alteration of even one of the other 34 destroys the functioning of molecules. Cytochrome C is an ideal molecule, it is relatively small and stable, is colorued and is ubiquitos in aerobic organisms. Its sequences in atleast 25 species of vascular plants have been determined so far. Besides others in algae, animals, fungi, and bacteria. The number of differences is closely parallel to the distance apart of the organisms in traditional classification, suggesting that the method is broadly reliable (Boulter, 1974, Boulter et al., 1972).

Doubtlessly amino acids have very restricted distribution the feature have been copiously exploited to characterise certain sub families and genera. In the sub family Lotoideae of Fabaceae Conavanine has been observed (Turner and Harborne, 1967). According Bell (1971) Lathyrine has been found to be restricted only to the genus *Lathyrus*.

Electrophoresis

Electrophoresis is one of the widely used techniques in protein investigations. Generally electrophoresis involves the migration of proteins across a voltage gradient. Disc electrophoresis (DEC) in particular is a form of zone electrophoresis where in proteins are completely separated within a gel medium. Charged molecules will move in a certain direction when subjected to a direct current. This method detect similarities and differences of protein molecules based on two specific properties which affect mobility, ionic charge and molecular weight. Disc electrophoresis derives its name from the fact that two discontinuous media are placed in conjunction, facilitating the sharp separation of proteins into narrow bands. The gels basic characteristic is a sieving effect which allows for high resolution separation of proteins. The gel is removed from the voltage gradient, and the separated proteins, enzymes, and / or carbohydrates are located by using selected gel. From the reported data it appears as if this method providing protein band data useful at various taxonomic levels. Vaughan et al (1966, 1967) have differentiated species of *Brassica*. Fox *et al* (1964) have indicated electrophoresis can be used at tribal grouping in the Leguminosae. Thurman *et al* (1969) compared the mobility values of three hydrogenases from taxa of the Leguminosae species and tribal levels. Vaughan *et al* (1966) compared the albumin and globulin fractions of the seeds of three Brassica species.

According to Johnson (1972), the cereals related to what in relation to the genomic constitution and ancestry of the tetraploids and hexaploids, the most valuable electrophoretic results have perhaps been obtained. For example Johnson (1972) Johnson and Hall (1965) investigating the storage proteins, concluded that hexaploid bread wheat *(Tricitum aestium)* indeed contain a sum of the proteins possessed by the diploid species. Based on the morphological and cytological evidence the diploid species which had been postulated to be ancestral to it.

Numerical Taxonomy

Heywood, (1967) defined numerical taxonomy as numerical evaluation of the approach (Phenetic) similarity among the groups of organisms and ordering them into higher ranking taxa. Numerical taxonomy does not produce new data. From various branches of botany such as morphology, Anatomy, Palynology, Cytology, the translated variation into numerals are processed through standard statistical notations and mathematical formulae to bring out significance of the character. The effective use of that character is evaluating taxa, or genus or family in the classification with theoretical background aims and principles far making comparison.

For classifying plants Michel Adonson (1763) for the first time pointed out that due weightage should be given for all the characters. These principles are called Adonsonion principles which gained importance since 1960s and have produced new methods in taxonomy includes under a general term 'Numerical Taxonomy'. The term may be defined as numerical evaluation of apparent variation character from various sub disciplines of botany such as morphology anatomy, palynology and cytology. Differences and also ordering them into higher taxa (Heywood, 1967). Numerical taxonomy is empharical science and aims to develop objective methods. The propositions can be tested through repeated experiments for evaluation of taxonomic relationships and in the erection of taxa in hierarchial system of taxonomy. In addition the numerical methods have provided clues to phylogenetic analysis for evolution of evolution rates more accurately.

The particular branch of botany concerned with numerical approaches towards taxonomy has been variously termed as Numerical Systematics (W.H. Wagner) Mathematical Taxonomy (Jardin and Subson, 1971) Taxi matrics (Mayr 1966) Taximetrics (Roger, 1963) and Numerical Taxonomy (Sneath and Soakal, 1973).

First of all the individuals are selected and there after the characters are specified in Numerical Taxonomy. Based on the character analysis the similarities amongst individuals are established with the help of the computers, the stages are often worked out. The estimation of the accuracy of similarities between taxonomic groups depends on the appropriateness in character. Choice of character requires a planful approach while working on these lines. Based on the existing literature or original observations from the specimens themselves, the selection of characters may be made. As far as the character number is concerned there is no limitation, but larger the number better is approach for generalization of the taxa. It is widely accepted a number of 80-100 character is desirable in numerical taxonomic studies.

There are two aspects of Numerical taxonomy the construction of taxonomic groups and their dissemination.

Construction of Taxonomic groups

For establishing natural system of classification, two major aspects have been recommended. Firstly phenetic and secondly phylogenetic (Cladistic).

Phenetic classification is the similarity (relationship) of the individual under consideration based on a set of a phenotypic (characters related to the appearances) characters. This phenotypic relationship expressed in terms of similarity does not show any relationship through ancestry.

In terms of correlation amongst individuals with regard to their evolutionary history the cladistic relationship is expressed. The relationship means common ancestry. The word cladistic means study of evolutionary sequence and pathways followed by individuals and the origin of branches from evolutionary tree. The phylogenetic tree with anastomosing branches leading to various taxa are termed as cladograms Operational Taxonomic Units (OTUs).

In numerical taxonomy the basic unit is known as operational taxonomic unit (OTU). It can be individual, species, genus, family order or class. When OTU is supra-individual (above the level of an individual) there should be adequate representation of various polymorphic forms i.e. when genera are compared they should be represented by different genera. When families are compared, they should be represented by different genera and so on; It should always be borne in mind that comparison of OTUs of equal rank is made in numerical taxonomy.

Unit Characters

The characters used in Numerical taxonomy are known as unit characters. A unit characters is one which cannot be sub divided logically. There are two types of unit characters.

a) Those unit character which exist in two states. They are called binary or two state characters. This is the simplest form of coding where characters are divided into + and - or as 1 and 0. The positive characters are recorded as + or 1 and negative characters as - or 0. In case the organ processing a character is missing in an organism the characters is the presence or absence of spines.

b) Those unit characters which exist in more than two states are called multistate characters. The multi state characters may be (i) Qualitative such as the color of the flower, the color of the flower could be in any number of states - white, red, yellow, violet, blue etc or (ii) qualitative such as length of the leaf, it could be in any number of states - 1cm, 2cm, 3cm, 4cm, etc. The multi state characters can be converted into two state characters like flowers white versus colored leaf long versus short. It may be observed that characters state an attribute that exist in abstract form. While the character state is an expression of the character that exists in concrete form. The phyllotaxy is a character and alternate opposite or whorled forms are characters states. There is another term characteristic associated with the above two terms. When a character state is exclusive to a particular taxon/taxa it is called characteristic. In the application of numerical taxonomy of the proper selection of characters form a critical Sokal and Sneeth (1963) disqualified certain characters for numerical taxonomy and these are precisely inadmissible characters.

According Sokal and Sneeth (1963) it is undesirable to use :

1) If the attributes which are not a reflection of the genotypes of the organisms themselves.
2) If any character which is a logical consequence of another either partly or wholly and
3) If the character which do not vary within the entire sample of organisms.

To be qualified as admissible character:

i) The character should be inherent in the organism and should neither be susceptible to environmental changes.
ii) The characters must be of some diagnostic value
iii) The character cannot be divided
iv) The selected character must show variation among the taxa under comparison.

Measurement of Resemblances

There are three methods of estimating phenetic resemblances between the taxonomic groups namely :

i) Coefficient of association
ii) Coefficient of correlation and
iii) Measurement of taxonomic distance

Cluster Analysis

Different OTUs are grouped together on the basis of degree of similarity. These groups of OTUs are termed clusters, there are several techniques to describe structure in matrices of similarity coefficients. The differential shading of the similarity matrix is one of the common techniques. In this method similarity coefficient are grouped into five to ten evenly spaced clusters. Each of these classes are represented by different degrees of shading in the square of half matrix. The higher value is generally shown darkest and the lowest value the highest as in Fig.14.3. Then half matrix can be seen as a pattern to different shades limited by diagonal squares with the darkest shades Fig. 14.3.

On rearrangements of the sequence of OTUs clusters can be more sharply defined as in Fig.14.4.

The similar organisms which are grouped in this manner are termed phenons. The clusters of pheonons are then rearranged in a dendrograms which summarises the main features of cluster analysis.

The second way of clustering is by dendrograms. In this case the mutually most similar taxa are paired. The pairs are successively joined by the average similarity. In this process all units have been joined together. It forms in a tree like dendrogram with taxa at the tips of branches.

The horizontal (Phenon line) interesting the vertical lines of dendrogram gives number of clusters.

The limitation of phenon is done by drawing a horizontal line across the dendorgram (Fig. 14.4) at a similar value. A line at 75% for example creates five 75 phenons 1, 7, 3, 5, 16, 4, 9, 10 and 2, 8 while at 80% creates six 80 phenons. Such dendrogram will have a reference to a given taxon and cannot be transferred to any other study. In the above dendrogram if the OTUs, 1 to 40 had been species on 80 phenon line could indicate 6 sub genera and a 65 phenon line the genera. It should be remembered that phenons are arbitrary and relative groups.

Phenons and rank

Sneath and Soakal (1962) have introduced the concepts of phenon for definition of groups obtained by cluster analysis.

They may be prefaced by number indicating the level (percentage) of similarity at which they are formed e.g., an 80- phenon indicates a group whose members are affiliated at no lower than 80 percent of the similarly scale used in the particular analysis. In practice a phenon defines a groups by drawing line horizontally across the dendrograms: these lines must be straight and parallel to abscissa. In this way similarity

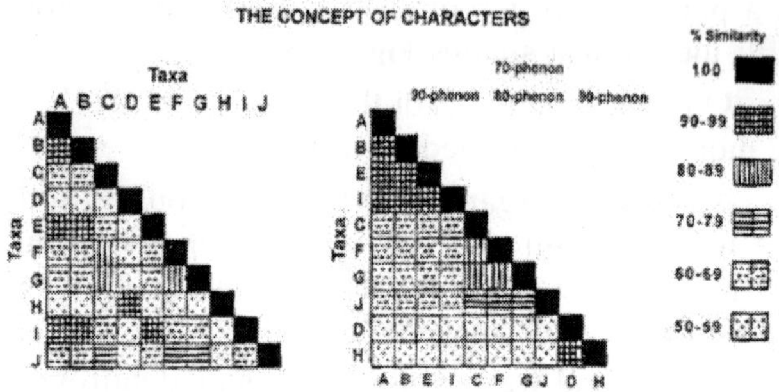

Fig 14.3 Shaded similarity matrices

A. Percentage similarity B. OTUs arranged haphazardly

C. After rearrangement of OTUs

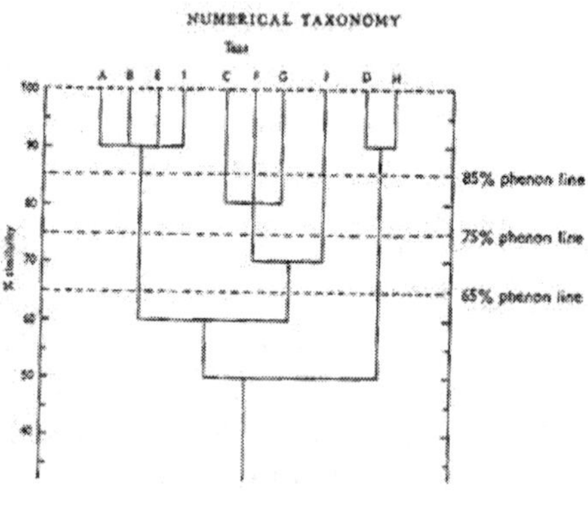

Fig. 14.4. Dendograms to show formulation of phenons

may be used as criteria of rank: we can thereby delimit taxonomic groups objectively by choosing orbitrarly fixed levels of similarity. For example in fig 14.4 by drawing a horizontal line across the dendrogram at a similarity level of 75 we have created a four 75 phenons. (ABEI, CFGJ, and DH), these may be regarded as subgenera, genera etc. depending on the rank of the original taxonomic units employed in the study clearly such methods must be applied uniformly through out a group under study phenons need not however, be equated with taxonomic groups unless so desired. The advantages of such methods for delimitation of supraspecific taxa by agreed orbitrary levels of similarity are obvious in theory at least. Methods for comparision of dendrograms derived by numerical methods with one another and with those produced by conventional means have been proposed by Sokal & Rohlf (1962). Models for the presentation of the major relationships of taxa as derived by cluster analysis have been used by Lysenko and Sneath (1959) can be verified by any one.

Discrimination

In the study of taxonomic groups chosen show overlapping characters, discrimination should be used to select them. Various techniques, such as discriminant analysis, have been divided for such purposes. The best methods for delimiting taxa are based on the utilization of maximum number of characters with similar weightage given to them.

Nomenclature and Numerical Taxonomy

Modern nomenclature does not concern itself with the problem of delimitation of taxa. It seems only as a reference point to the taxonomic names. The limits are debatable subjective and forever changeable. Numerical taxonomy on the other hand is very useful in delimitation of taxa by exact estimation of affinities (although phenotic). Thus there is no scope for personal opinion or decision of taxonomist. The limit may be objective utilitarian permanent and fixed by common consent. It is most unlikely that numerical taxonomy will over supplant traditional methods as the standard procedure. For the foreseable future it is likely to be most useful in examining classifications, assessing their good and bad features, and in some cases pointing ways of modifying and improving them. In phylogenetic research, where by numerical techniques are used to determine the minimum length branching dendrograms that it is necessary to postulate in order to arrive at the known phenetic pattern of taxa is most likely phylogenetic route.

REFERENCES

Davis, P.H. and Heywood, V.H. 1963. Principles of Angiosperm Taxonomy. Robert.E. Krieger Publishing Company, New York, London.

Stace, A. clive, 1984. Plant Taxonomy and Biosystematics Edward Arnold (Publishers Limited, London).

Pullaiah, T., 1997. Taxonomy of Angiosperm. Regency Publications, New Delhi.

Radhakrishnaiah, M., 1996. Essentials of Plant Taxonomy Self Publishing.

Singh.V and Jain, D.K. 1989. Taxonomy of Angiosperms Rastogi Publications, Shivaji Road, Meerut.

Sneath, P.H.A., Soakal, R.R. (1973). Numerical Taxonomy, W.H. Freeman, San Fransico.

15

Self Incompatibility

Objectives

1. To assess the angiosperms plants opting for cross pollination and avoiding self pollination
2. To assess and study the architecture of flower tend to develop the heterostyly, Dichogamy (protandry and protogyny)
3. To assess and study the unisexuality and dioecism and gynodioecism
4. To understand the evolutionary significance of incompatibility.

Introduction and Definition

Self incompatibility is a mechanical devices of the flower to prevent self pollination (Synanthesis) and temporal features having a similar effect or dioecium and self incompatibility. Inability of the plant producing functional gametes to set seed upon self pollination (BrewBaker, 1957) Landquist, (1964) and de Neltomcourt (1977) defined it as the inability of a fertile hermaphrodite seed plant to pollinate to produce zygote after self pollination. In the crop, perennial grasses, or ornamental plants, and forage, legumes and members of Solanaceae, Rosaceae, Asteraceae, Brassicaceae have S1 mechanisms of different kinds and degree of effectiveness. Those self incompatibility mechanisms include the following.

List of Mechanisms involved in Self Incompatibility

1) The pollen grain fails to germinate on the stigma is self incompatibility.
2) The pollen tube fails to enter or penetrate the stigma even though the pollen grain germinates.
3) The flower may drop down before the fertilization (the pollen tube may penetrate into the style).
4) Even the pollen tube reach the ovule but fail to fertilize the egg.
5) Even if fertilization is over but embryo degenerates at an early stage.
6) Due to some chemical repulsion the fertilization does not occur
7) Due to difference in maturity of pollen or ovule fertilization fails since the difference in things they became sterile.
8) Self incompatibility impedes self pollination and thus promotes cross pollination.

Promotion of Cross Pollination by Appressing Self Pollination

Control at the level of pollen germination and egg fertilization

i) Multigenic systems: S type – sporophytically determined incompatibility and gametophytically determined incompatibility.
ii) Diallelic systems: with associated morphological differences (heterostyly) and without associated morphological differences.

Control in timing of pollen maturity and availability of mature styles

protandry and protogyny

Control in the physical availability of receptive styles

i) Unisexuality - monoecious decline, dioecious and polygamodioecious

ii) Architecture of the flower operating mechanically to impede selfing.

Mechanisms innovated in Flower to Avoid Self Pollination

Charles Darwin in his book *"The effect of cross and self fertilization in the vegetable kingdom"* (1888) describes the following experiments. He conducted pollination experiments with *Ipomea purpurea* (*Convolvulus major*). Ten flowers were self fertilized and ten flowers from the same plant were fertilized with pollen from a different plant. Darwin continued the experiments for ten generations, crossing the progeny of the time originally crossed and selfing that of the originally selfed. The same advantage in heighest number of capsules, number of seeds per capsule and weight of the crossed over the self persisted but measured in magnitude so that by the end of the third generation. The crossed plants produced seeds in comparison with self-fertilized plants in a proportion of 100 to 35, the cross fertilized seeds weighing (43.27) grams / 100 seeds and the selfed 37.63 / 100 seeds.

Darwin repeated this kind of experiments with several dozen cultivated and wild plants, obtaining similar results, but in certain cases the self fertilized and the cross-fertilized plants showed no appreciable difference. One such case is *Mentzelia aurea* (*Bartonia aurea* in Darwin's account) in the Loasaceae. Nevertheless, the plant did not grow well. Data on the plant height could be obtained and not on seed set. In another species, crossed plant did not show any advantage over the selfed in Pisum sativum and Darwin suspected genetically identical plants resulting from previous constant inbreeding.

Generalization of Darwins Experiments

Cross fertilization is beneficial and self fertilization injurious. Later experiments in these lines corroborated

Darwin's insight. If the pollen incapable of germinating on the style of flowers of the plant that produced it or incapable of fertilizing the egg a condition known as physiological or genetic self incompatibility.

Multigenic systems: the most common type of self incompatibility is the multiple allelic kind found in each major phyletic line of the flowering plants including over eighty angiosperm families, gymnosperms and ferns. (East, 1940, Bateman, 1952, Fryxell, 1957).

This type of inheritance of self incompatibility was worked out in the genus *Nicotiana* by East and Mangelsdorff, 1925. In this sort of plants, plants fail to set seed when self pollinated, though pollen and eggs are fully fertile and produce seed abundantly. When these plants are crossed reciprocally with other individuals of the population. In these plants genetic analysis exhibited that each plant is heterozygous for a self incompatibility factors. A series of alleles in these population indicated as S^1, S^2, S^3, S^4, S^5-S^{11}. The pollen tube will not grow in the style that possess the particular S allele present in the pollen grain itself. Thus in the styles of a plant with S^1 / S^2 neither S^1 and S^2 pollen grains grow well regardless of whether produced by the same plant nor by another one while a pollen grain with any other S allele grow perfectly well. (Fig 15.1.)

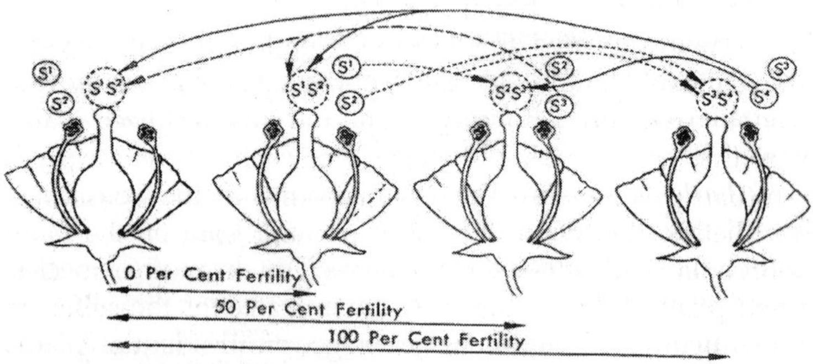

Fig. 15.1. Diagram showing the crossing relationships between self-incompatible plants. Unbroken circles show pollen grains; broken circles show genotypes of styles. (From Solbrig, 1966a.)

Multiple Allelic Incompatibility

Plants of self incompatibility possessing a multiple allelic incompatibility system can be divided into two groups: 1) the gametic type and 2) the sporophyte type. In the gametophytic type each pollen grain expresses the incompatibility type of the S gene it carries whereas in the sporophyte type, the pollen grain also expresses the incompatibility reaction of the S genes of the maternal plant. The genetic composition of the pollen determines the incompatibility reaction whereas the genetic composition of the tissue of plants determines the incompatibility reaction in the second instance.

The cytological, physiological and nuclear aspects in the self incompatability not worked out fully. Sporophytic type of self incompatability (See table 15.0) are associated with trinucleate pollen and simultaneous formation of cell walls at meiosis though species with heteromorphic flowers have sporophytic incompatibility reactions and binucleate pollen. According to Pandey (1960) the sporophytic type of incompatibility is derived from the gametophytic type and

TABLE 15.0

Relationship Between Type of Incompatibility, Type of Pollen Mother Cell Division in Meiosis, and Number of Nuclei in Pollen Grains.

(Adapted from Pandey, 1960)

Flowers Homo-Morphic or Heteromorphic	*Type of Cytokinesis in Pollen Mother cell*	*Number of Nuclei in Pollen Grains*	*System of Self-incompatibility*	*Number of Families*
Homomorphic	Both simultaneous and successive	2	Gametophytic	2
Homomorphic	simultaneous	2	Gametophytic	28
Homomorphic	successive*	2	Gametophytic	5
Homomorphic	successive*	3	Gametophytic	1
Homomorphic	simultaneous*	3	Sporophytic	8
Heteromorphic	simultaneous*	2	Sporophytic	5
Heteromorphic	simultaneous	3	Sporophytic	4

* *In some cases a family may have some species with simultaneous and others with successive cell division. The incompatibility depends in such cases on the particular kind of cytokinesis the species has, in accordance with the above table.*

functionally is related to an earlier time of action of the S allele so that all four products of meiosis receive the substance or substance that are involved in the incompatibility reactions. In gametophytic type of incompatibility, gene action would not occur until the first cell wall after the first metaphase division of meiosis was formed and the substances produced by the gene could not diffuse freely through out the four cells (Fig 15.2).

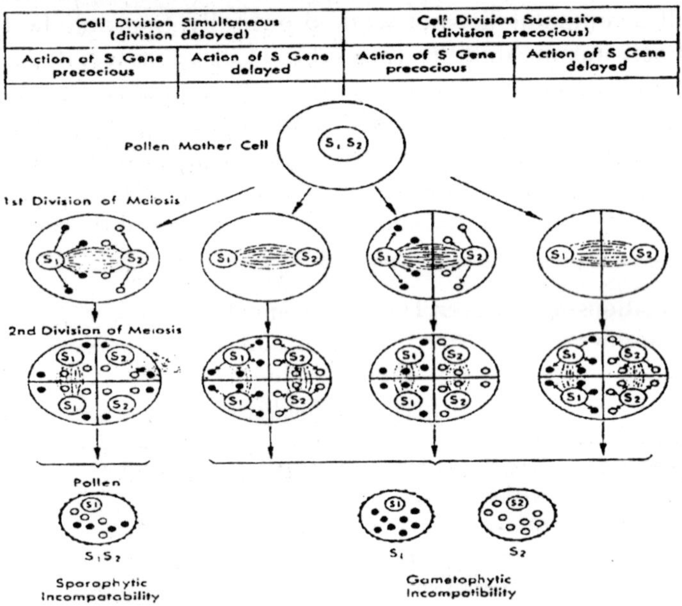

Fig. 15.2. Pandey's hypothesis (Pandey, 1960) to explain the correlation between cytological characteristics and type of incompatibilities. According to this hypothesis, the incompatibility genes produce a substance (represented here by black and white dots) that control the incompatibility reaction the pollen grain. In the case where the action of the S gene is precocious (that is, it takes place before the onset of the second meiotic division) and cell division is retarded (taking place after the information of the four

nuclei of the tetrad, the so-called simultaneous type), the substances of both genes are evenly distributed throughout the cytoplasm, and the pollen grain has a sporophytic type incompatibility. In all other cases the incompatibility reaction is gametophytically determined.

Heterostyly and Self Incompatibility Mechanism

The existence and evolutionary role of heterostyly was first explained by Darwin in a series of articles between 1862 and 1868 and extended in his book "The different forms of flowers on plants of the same species (1877). Darwin observed, in the common garden primroses, *Primula vulgaris* two kinds of flowers on different plants. In one kind a long style and short stamen, in the other long stamens and a short style (Fig 15.3). Darwin crossed the long styled flowers (known as pins) with other long styled flowers and also with short styled flowers (known as thrums) and these last with other thrums as well. The crosses between thrums and pins yielded a large seed crop, while crosses between the same types of flowers yielded little or no seed. Heterostyly usually involves differences in the length of the anther and style, some times two kinds of such as in *Primula* or less frequently three kinds as found in the genus *Oxalis*.

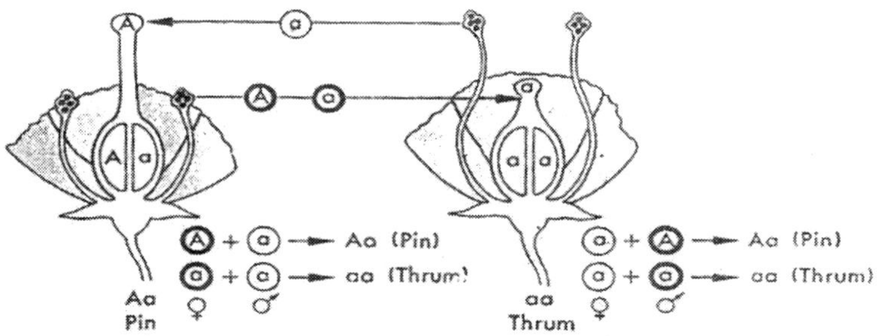

Fig. 15.3. Diagram of the crossing relationships between heterostylous (in this case distylous) plants. An a-pollen produced by the pin plant reacts on the thrum style as if it were A. Note that both pin and thrum plants yield exactly one half pins and one half thrums in their progeny. (From Solbrig, 1966a.)

Incompatibility systems favours outbreeding. They depend, as dioecium on genotype differences but they operate in hermaphrodite flowers. Self incompatibility is a breeding system in hermaphrodite organisms which prevents the formation of zygote after self mating (Lewis,1954). It is wide spread amongst Angiosperms and is known from more than 3,000 species in some 20 families and is probably many more examples remain to be discovered. Few species have been well analysed largely because of the labour involved.

There are different degrees and types of incompatibility. A distinction is made between self-incompatibility which operates between pollen grain and style of the same plant (i.e., prevents the self fertilization of hermophrodite) and cross-incompatibilty which operates between the pollen and style of different individuals and races.

Three main systems of self incompatibility are found in the Angiosperms. The incompatibility reaction may be between 1) pollen grain and stigma, 2) pollen tube and style or 3) between the male and female generative nuclei in the embryosac.

(1) Heteromorphic: (e.g. Primulaceae, Lythraceae, Linaceae, Rubiaceae)

Here there are two or three morphologically distinct mating groups: long and short style flowers in distyly, long, medium and short style flowers in tristyly. The pollen reaction is sporophytically controlled. The plant controls the characteristics of the pollen by means of morphologically differentiation of the flower (Darlington and Mather, 1949). Genetic control is by one gene with two alleles S and s is distyly or by two genes each with two alleles Ss and Mm in tristyly.

(2) Heteromorphic, gametophytic (Solanaceae, Rosaceae, Leguminosae, Scropulariaceae). Here all the mating groups are indistinguishable morphologically. The pollen reaction is gametophytically controlled. Genetic control is by one gene with numerous alleles $S_{1, 2, 3 \ldots n}$ (multiple allelomorphic incompatibility).

Homomorphic sporophytic (Compositae, Cruciferae, Rubiaceae): This system combines features of the other two systems- genetic control is by one gene with many alleles but the pollen reaction is under sporophytic control.

Detailed discussion of self incompatibility are given by Bateman (1952) Lewis (1954 a,b) and Pandey (1960) whitehouse (1950) has produced evidence to suggest that multiple - allelomorph incompatibility of pollen and carpel tissue may well have been primary cause of evolution of the closed carpel, and of the success which the Angiosperms achieved over their Gymnosperm ancestors. Further the implications of the self incompatibility in Angiosperms was well discussed by Bateman (1954) and Lewis & Crowe (1958).

Heterostyly

There are various accessory characters involved in the heteromorphic systems. In the classic case of *Primula* e.g. *P. sinensis*, in the "pin" and "thrum" flowers morphological features are, style length and stigma position, anther height and position, size and configuration of the pollen grains, size and configuration of the stigmatic papillae.

Such features are controlled by the S gene complex as in the Plumbaginaceae where the incompatibility system is allied to that of most heterostylous taxa, here there is a striking pollen dimorphism and in some cases an equally striking stigma dimorphism. Plants with the different kinds of pollen are found in populations of dimorphic species and those bearing each pollen type are self incompatible but cross compatible. In *Limonium vulgare* there is also stigma dimorphism with cob and pistillate stigmas corresponding to the two kinds of pollen. There is also heterostyly with long cob styles and short papillate styles. This culminates an evolutionary sequence in the tribe Staticaceae, derived from a primitive condition in which there is monomorphism has arisen in *Armeria* and *Limonium* this indicates self incompatibility, but the plants are still capable of some degree of outbreeding (Baker 1953,1954).

In *Linum grandiflorum* only the style differs in length. It is shorter or equal in length to the stamens. Tristyly is known in *Lythrum, Oxalis, Narcissus*. The mechanism in the latter is not strikingly speaking heterostyly, however since the variation is not apparently related to the breeding group. It is known that in *Narcissus cernuus* only single posture type may be found in a given area.

TABLE 15.1
Heterostylous families
(Adapted from Vuilleumier, 1967)

Family	*Type of Heterostyly*
Polygonaceae	D
Capparaceae	D
Saxifragaceae	D
Connaraceae	D
Leguminosae	D
Oxalidaceae	D,T
Linaceae	D
Erythroxylaceae	D
Guttiferae	D
Turneraceae	D
Lythraceae	D,T
Primulaceae	D
Plumbaginaceae	D
Oleaceae	D
Loganiaceae	D
Gentianaceae	D
Polemoniaceae	D
Boraginaceae	D
Rubiaceae	D,T
Commelinaceae	D
Pontederiaceae	D,T
Iridaceae	D
Amaryllidaceae	D

D= dimorphic; T= trimorphic

In *Primula vulgaris*, Crosby (1949) has studied the change over from out breeding to inbreeding associated with a change from heterostyly to homostyly. The significance of this example is that it can be studied in the act it has probably occurred and been completed in other species of *Primula*. Monomorphic species of the sections *Candelabra* and *Farinosae* are found near the edges of the range distribution suggesting their derivation from the main body of heteromorphic species, and monomorphic forms of normally dimorphic species of the sect. *Farinosae* have also been found at the margin of the ranges of these species (Ernst, 1953).

Other Diallelic Systems

Capsella bursa-pastoris (Riley, 1936), the sheperd's purse, possess a diallelic incompatibility system. Only two genetical forms are found, the homozygous recessive ss and the heterozygous Ss. S pollen germinates normally only on ss styles, while s pollen germinates normally only on Ss styles. In this respect *Capsella bursa-pastoris* resembles heterostylous species. Nevertheless morphological special characteristic of any kind has been associated with Ss or ss plants. This is so far the only non heterostylous diallelic system known.

Protandry and Protogyny

Both these mechanisms involve the different maturation of the organs. A plant is said to be protandrous, if the time of the release of the pollen, earlier to the gynoecium and gynoecium not yet developed to the point, where fertilization of the ovule is possible.

A plant is said to be protogynous, if at the time when the gynoecium is ready for the fertilization of the ovules, the pollen has not yet been released. Since blooming in a population is always staggered to a smaller or larger extent, in a protandrous species the styles will tend to be pollinated by the pollen of flowers that emerge a little later; the converse is true for protogynous species. However the time gap between the maturation of pollen and style can be very short; so that some pollen is available for self pollination towards the end of the time the flower stays open provided that no previous cross fertilization has taken place.

Protandry and protogyny are to be expected in species that are genetically self compatible, a mechanism to restore some degree of self incompatibility to self compatible species. Protandry and protogyny are know, however also in species that are self incompatible, as in many species of the family compositae.

Unisexuality, Dioecism and Gynodioecism

In some species of Angiosperms, the flowers are unisexual and two types those that produce only stamens and those that produce only carpels. If the two kinds of flowers are produced on the same plant, the individual is monoecious, whereas in the different kinds of flowers are produced on different individuals the plant is dioecious. In some other cases such as European ash, *Fraxinus excelsior*, male, female and hermaphrodite flowers are produced, such species are polygamous; Polygamodieocious if the different kinds of flowers are produced on different plants. Polygamomonoecious if they are produced on the same plant. Still other species, such as *Plantago lanceolata*, have only two kinds of flowers. Female and hermaphrodite flowers; such species are known as gynodioecious plants; if the two kinds of flowers are associated with different plants and gynomonoecious if on the same plant.

Unisexuality is another mechanism that prevents self fertilization. Unisexuality is clearly derived in most angiosperms, most unisexual flowers have parts of other sex in rudimentary form, such as in *Baccharis*. But in some species in which no rudiments of other sex are present such as willow.

Dioecism

Where the sexes are separate on different plants, is the most drastic system of out breeding, since not only is out crossing obligate but it involves the very considerable risk that fertilization may not be realized at all (Baker, 1959). There is, for example, a strong possibility that in any locality the neighbouring plants in populations will be of the same sex. Furthermore only about half the plants will be pistillate and therefore potentially capable of fertilization. Dioecism like monoecism may be incomplete; some species are polygamous (gynodioecious) where hermaphrodite and female individuals also occur, they may also both occur on the same plant. Gynodioecism is an out breeding mechanism, since the hermaphrodite flower may be selfed but the female flowers must be out crossed, the different mating types contributing unequal number of genes to the next generation (Lewis & Crowe, 1956). Gynodioecism is frequent in many genera of the Labiatae such as *Oreganum, Salvia, Mentha, Calamintha, Glechoma* etc.

It is also known in a number of other families including the Compositae, Caryophyllaceae and Saxifragaceae. Detailed studies on the gynodioecious breeding system of species of *Pimelea* (*Thymeleaceae*) have been made by Burrows (1960). High proportion of females were found in populations which ensured out crossing a low fruit set that was found in female flowers. Widespread hybridization in the genus is attributed by Burrows to the gynodioecious condition in many species. In the Lamiaceae the female sex flowers are typically smaller than hermaphrodite flowers; this difference may be mistaken for simple size variation and in several cases the sexual nature of polymorphism has not been appreciated. In the genus *Ziziphora* showed that sexual differentiation is quite common. There the length of the corolla tube, its degree of exertion from the calyx and degree of exertion of the stamens depend on the sexuality of the plant (Hedge, 1961). The corolla tube length has been used by previous taxonomist as key character in separating species of *Ziziphora*.

Gynodioecism should not be confused with the rare mutants. Causing male sterility in many hermaphrodite species. This is manifested in the abortion of pollen, but has little effect on outbreeding system of a population.

Within a genus both monoclinous and dioecious species may occur, for example in *Mussandra* where *M. tristigmatica, M. cheppii* are functionally dioecious while *M. erythrophylla* is heterostylous.

Sex Chromosomes in Plants

The unisexuality in animals is usually determined by a pair of heteromorphic chromosomes (the XY and ZW systems). Botanist have made much effort in searching for such chromosomes, the search has been quite fruitless. In some thirteen families of angiosperms where the genetics of sex determination is fairly well understood (Westergaard, 1958) there are some species with true sex chromosomes. In *Melandrium album*, one of the campions in the pink family, a European plant that has spread to many parts of the world as cultivated plant as a roadside weed. Plants of *Rumex hastatulus* found in Carolinas and south to Florida and west to Texas, Arkansas and Missouri and its populations have ten chromosomes. Female plants are homogametic i.e. they produce only one kind of gamete which has four autosomes and one X chromosomes. The males are heterogametic i.e. they produce two kinds of gametes each with four autosomes and either one X chromosomes or one Y chromosome but in North Carolina have eight chromosomes. The females have six autosomes and two X chromosomes, whereas the males have six autosomes one X and Y (Y_1 and Y_2) chromosomes (Fig 15.4).

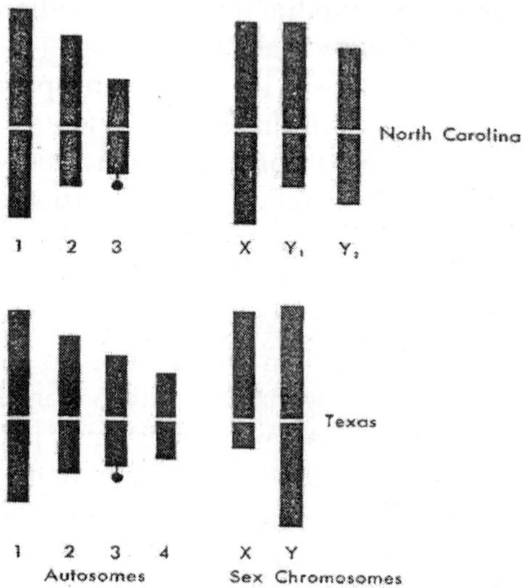

Fig. 15.4. Ideogram of chromosomes of a male plant of *Rumex hastatulus* from North Carolina (above) with three autosomes, one X and two Y chromosomes; and Texas (below), with four autosomes, one X and one Y. (From B. Smith, 1964, by permission.)

Ben Smith (1964) has shown that the changes are due to a series of translocations. Fig 15.5. He postulates that selective force favoring the change is the lower recombination index that a low chromosome number presumably gives.

COORIENTATION OF SEX CHROMOSOMES AT MEIOSIS

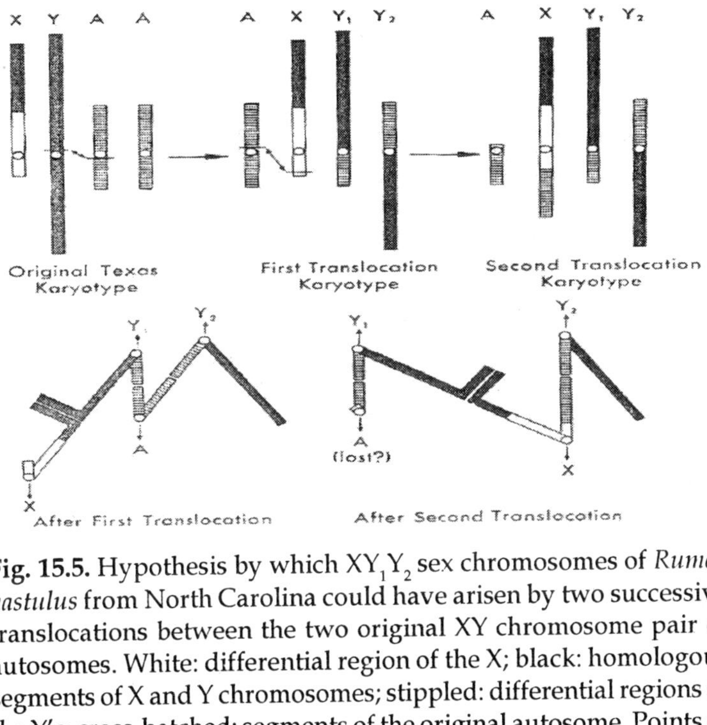

Fig. 15.5. Hypothesis by which XY_1Y_2 sex chromosomes of *Rumex hastulus* from North Carolina could have arisen by two successive translocations between the two original XY chromosome pair of autosomes. White: differential region of the X; black: homologous segments of X and Y chromosomes; stippled: differential regions of the Y's; cross-hatched: segments of the original autosome. Points of assumed breakage and translocation are indicated by double-headed arrows. (From B. Smith, 1964, by permission.)

The situation in the campion *Melandrium album* does not involve changes but in the proportion of the two sexes in the population. In species with an xy chromosome sex determining mechanism was amplified i.e., in *Melandrium album*, approximately equal numbers of both sexes are expected in the population, because 50% of the pollen will produce male plants after fertilization and 50% will produce female plants.

Evolutionary Sequence in Leavenwarthia Flower

A very fine example of reversion to self pollination is known in the genus *Leavenwarthia*. Of the eleven species found in the genus, three species: *Leavenwarthia alabamica, L. stylosa* and *L. crassa* are primarily self incompatible; all the others are self compatible. The change in breeding system has been accompanied by a series of morphological changes to assure self-pollination (Fig 15.6). In the self compatible species the flower have become smaller, the styles shorter so that they are level with anthers, and anthers now open to the inside rather than outward.

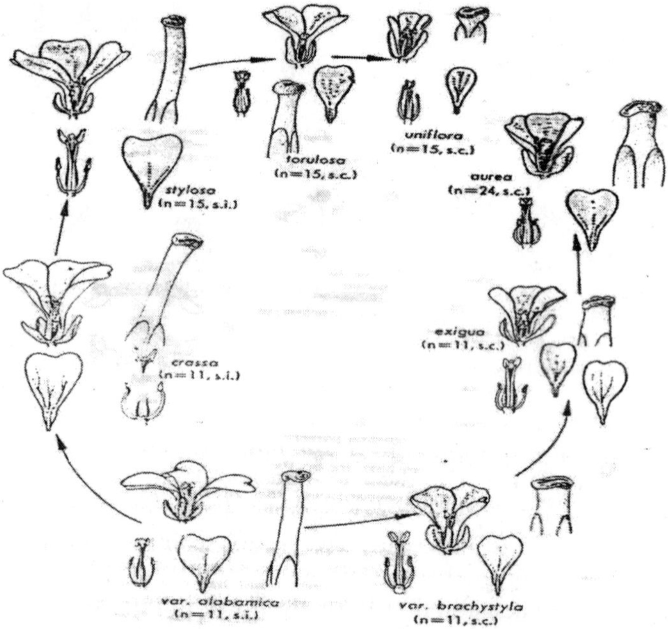

Fig. 15.6. Flower, stigma, stamens, and petals of species of *Leavenworthia* arranged in their probable evolutionary sequences, starting with *L. alabamica*. Note how the flowers become smaller, the styles become shorter, but not the stigmas, and how the stamens tend to grow appressed to the styles, with increasing self-compatibility. Flowers and petals two times natural size, stamens three times, styles, ten times. (From Rollins, 1963.)

Evolutionary Significance

Darwin from his work on the effect of cross and self fertilization in the vegetable kingdom indicated it was nature's urge producing vigour and maintaining fertility. That is out breeding systems of various kinds result in gene recombinations which play as great a role as mutations in providing the variation on which natural selection works. The gene recombinations in turn lead to the new combinations of characters on which natural selection may operate to produce in time better adapted structures and organisms. Selection may be *allaesthetic* as in the case of the genus *Ophrys* where various flower types which arose randomly as a result of mutation and recombination in the ancestral population were selectively perpetuated in the assertive pollinations (pseudocopulation) and these selected types were then perfected to the stage which we know them today (Stebbins, 1959; Heslop-Harrison, 1958). The out breeding population a reservoir of recessive genes held in heterozygous condition. The hidden variability is of considerable significance, since it will produce an array of genotypes some of which may be adapted to meet new conditions which the species might come to face as in changing environment.

According Lewis & Crowe (1958) an organism when competing in a common environment with other organisms. At this stage out breeding mechanisms are essential for survival by giving advantages of gene recombination.

According Gause's principle (1934) that two or more forms with identical requirements cannot coexist in the same environment for an indefinite period of time. After establishing itself in an environment, the out breeding system may be retained or for some reason abandoned.

In some instances such races have been included in the same species until a knowledge of breeding system has been obtained. Baker quotes (1961) the genus *Primula* where in the 19th century several geographically and morphologically distinct taxa of subsect. *Farinosae* were included in *P. farinosa*. Later *P. farinosa* sensu stricto is diploid, heterostylous and largely self incompatible, while closely related forms constitute a chain of homostylous polyploids which must show genetic isolation from their diploid relatives. In the light of this knowledge there is no alternative to considering them separate species.

According to Baker (1959, 1961) all these systems promoting outbreeding are reversible, permitting inbreeding. Thus morphological changes may take place in the flowers. Self incompatability alleles may arise in an incompatibility system, heterostyly give rise to homostyly (e.g. of *Primula*) etc. The reverse evolution from inbreeding to out breeding is apparently much rearer. This flexibility is of a great evolutionary advantage, permitting adaptations to changing condition. Apomictic groups, on the other hand are correspondingly limited.

16

THE ROLE OF HYBRIDIZATION AND EVOLUTION IN TAXONOMY

Objectives

1. To understand the three ways of classification of evolutionary process in Hybridization.
2. To understand the survival of Hybrids in the disturbed or intermediate habitat and eroded areas.
3. To assess how the introgressive hybridization stabilize the hybrid survival.
4. To assess how the role of hybridization via introgression lead to stabilization.
5. How the barrier reproductive isolation build up in hybridization.

Definitions

Hybridizations normally implies the known or assumed crossing between plants belonging to different Taxa, usually species, sub-species or in rare cases genera. The hybridization for evolutionary purposes is considered by Stebbins (1959a) to be as follows; crossing between individuals belonging to separate populations which have different adaptive norms as also defined by Mayr (1942); the hybridization between reproductively isolated species. In so far as the Hybridizing population can be recognized as different Taxa. The term Hybridization to crosses between plants presumed to belong to different species. For crosses with in the same species the term recombination is used.

The most obvious effect of Hybridization is to break down and blur the distinctions between species.

Three Ways of Classification of Evolutionary Process in Hybridization

Stebbins (1959) in his important review of the role of Hybridization in Evolution, classifies the evolutionary process involved under three main headings. (i) Stabilization of Hybrid derivatives by means of amphidiploidy. (ii) Stabilization by introgression (iii) Stabilization of segregates from hybridization. The introgression and segregation may result in non-stabilization, instead of stabilization and in every transition between them. The fact the most hybrids between distinct taxa show some kind of loss of fertility indicates that a certain amount of genetic isolation accompanies morphological and physiological divergence.

Role of Hybridization in Evolution

Whenever two species that possess no genetic sterility barrier are within pollination range of each other, hybridization can occur. Pollination range is defined as the distance within which most of the pollen of species is dispersed. The hybrids that are formed usually combine the characteristic features of both species morphologically physiologically and ecologically.

Survival of Hybrids in Intermediate or Disturbed Habitat

This means that if the parents have different requirements and such as for water, soil, light or any other characteristic of the environment the hybrid will probably not able to grow successfully in the environment of either parents. However in the area where the habitats occupied by the two parental populations come into contact, the physical habitat is often intermediate. Here the hybrids may successfully become established. Perhaps more commonly the hybrids become established in areas of disturbance such as clearings, erosion, gullies, cliffs and so on. In areas of disturbance the conditions required by the parental species are not met in their entirety; but the some what modified and intermediate requirements of the hybrids often make them successful inhabitants of such areas. Man in recent years become a major source of habitat disturber; but phenomenon such as glaciations, mountain building and erosion have created disturbed habitats in the geologic past.

Back crosses of hybrids with parents

The evolutionary role of hybridization depends on the effect that hybridization has on the genetic composition of the parental populations and not necessarily on the frequency with which they are formed. In most cases of hybridization it is not the hybrids but the backcrosses of hybrids with the parents that play an evolutionary role. This process of back crossing is known as introgressive hybridization, Anderson (Anderson, 1949). By introgressive hybridization, genes from one species can be introduced into populations of another.

Introgressive Hybridization

The effect of hybridization has on the genetic composition of parental populations and not necessarily on the frequency with which they are formed will be reflected on the evolutionary role. It is not the hybrids but the backcrosses of the hybrids with the parents that play an evolutionary role. This process of backcrossing is known as introgressive hybridization (Anderson, 1949). By introgressive hybridization genes from one species can be introduced into populations of another species and Anderson stated further in introgressive hybridization the backcrosses are favored by natural selection because of the new genes they carry.

A breeding population and their individual plants are more or less closely adapted to the environment in which they grow. The mutations and recombinations will produce individuals with genotypes closely resembling the parents and also individuals differing from them more or less in each generation. In a stable environment most of the seedlings that survive to maturity will have a genotype and phenotype similar to the parental generation; in a changing environment the offspring will not resemble them quite much. The degree to which the offspring generation will differ from the parental generation depends on the degree of recombination (and all the factors that control recombinations) and the intensity of selection as has already been explained.

The greater the genetic recombination, the higher the probability that offspring will not resemble their parents; the stronger the intensity of selection, the more genetic change will occur. If a selection is stabilizing the changes will be minimal and will be largely genotypic; if selection is directional, the changes will probably of greater magnitude and will be reflected both in the genotype and in the phenotype.

Hybridization introduces new genes into a population and consequently increases the probability of genetic change. It acts as a factor that increases variability.

The Role of Hybridization and Introgression

1) An excellent example of the role hybridization in evolution is that of sunflower studied by Heiser (Anderson, 1952; Heiser 1961). The genus of American plants comprises four more or less distinct groups (Heiser et.al 1962) eleven or twelve species of tap rooted desert perennials. They are diploid with seventeen pair of chromosomes and are formed only in North America.

2) Forty more North American species of perennial plants that grew from crown buds, rhizomes or tubers and are either diploids, tetraploid or hexaploid.

3) Seventeen shrubby species from South America, two of them diploids, the others unknown cytologically.
4) Seventeen species of North American are annuals. There is now a positive evidence that of the fifty North American species of *Helianthus*, twenty five are involved in hybridization with one or frequently several other species (Heiser, 1961). The crosses seem to be confined largely to species within each of the groups.

The most common annual species are especially interesting, particularly the common sunflower (*Helianthus annuus*). The species is found in central and western North America from southern Canada to north Mexico. Morphologically the common sunflower is quite variable, not only within but especially between populations. Heiser (1954) subdivided the species into three subspecies and one variety: subspecies *lenticularis*, subspecies *texanus*, subspecies *annuus*, which includes the cultivated sunflower, variety *macrocarpus*, subspecies *annuus* is the ruderal or weedy sunflower common in the middle western United States, particularly in rail roads and vacant lots in cities and towns; subspecies *lenticularis*, on the other hand, is the wild sunflower of roadsides and abandoned fields in the West and Southwest of the United States. Finally subspecies *texanus* is restricted to eastern Texas. The story of the probable origin of these three forms is a fine example of the role of hybridization in evolution.

Gradual development of annuus sub species:

The ancestor form of sunflower is not known with certainity. However on the basis of extensive studies Heiser(1954) postulates that subspecies *lenticularis* of western North America is more likely the original form of the species than any other living subspecies. This subspecies, according to Heiser gave rise to subspecies *annuus* under the indirect influence of man. The seeds of subspecies *lenticularis* were gathered for food by various tribes of Indians (Heiser, 1951), and the sunflower may have become an early Indian camp weed adapted to disturbed areas around the camps and villages. In time and either through the conscious or unconscious selection by man subspecies *lenticularis* could have given rise to the slightly larger forms that developed into subspecies *annuus* differs basically from variety *annuus* by single mutation that restricts, branching and leads to the production of a single or a very few large heads. Man has of course in addition selected forms for higher yield, especially in recent times.

To trace the history of the third species, *H. annuus* subspecies *texanus* is more involved. It appears to have arisen after the introduction by either man or natural causes of either subspecies *lenticularis* or *annuus* into Texas. Today *H. annuus* subspecies *texanus* is very common on road sides and abandoned fields, especially in central Texas. In eastern Texas grows another annual species of sunflower, *Helianthus debilis* variety *cucumerifolius*, found both as a roadside used and in undisturbed sites where *Helianthus debilis* variety *cucumerifolius* and *H. annuus* subspecies *texanus* grow together, occasionally a few hybrid plants are found. These are more stunted in growth than the parental species that grow in the variety and are highly sterile, although they produce a very small production of viable pollen. The variation of *Helianthus annuus* in eastern Texas (Table 16.0) is clearly in the direction of *H. debilis* (Heiser, 1951) in regard to the large proportion of leaves with jagged serration, speckled stems, small head diameters, small achenes, low number of ray flowers, small bract width and the tendency to branch from the base. Virtually it has been postulated (Heiser, 1951) that variety *texanus* arose as a result of hybridization and introgression of *H. debilis* variety *cucumerifolius* with *H. annuus* subspecies *lenticularis* and subspecies *annuus* when this last species was introduced into Texas. In addition to the morphological characters that *H. annuus* acquired from *H. debilis*, it presumably also acquired physiological characteristics that adapted it to withstand better the climate and environment of Texas. Characteristics acquired by *H.debilis* over many years of existence in Texas.

Table 16.0

Species	Locality	No. of ray florets	Disc diameter(cm)
H.debilis var. Cucumerifolius	Nacogdoches co.Texas	13.7	1.62
	Frio co.Texas	15.7	1.84
	Houston co.Texas	19.0	2.24
H.annuus	Galveston co.Texas	17.3	2.30
	Frio co.Texas	20.4	2.58
	Galveston co.Texas	22.0	3.28
	Collin co.Texas	23.2	3.60
	Pittsburgh co.Okla	25.9	3.54

In California, at the other end of the distribution of *Helianthus annuus*, 1) A similar phenomenon of hybridization can be observed. The sunflower is relatively recent resident of California. For the sake of seeds as food, it was probably introduced first by the Indians who used the flowers in their religious rites. In the beginning it was restricted in its distribution to disturbed sites likely around Indian camps. For the last hundred years or so as a result of activities of European man, disturbed sites and have multiplied enormously in California, and so has the Sunflower which is now very widespread along roadsides and railroad tracks and as a weed in cultivated fields.

Development of sub species annuus from H. bolanderi through back crosses or in disturbed areas

Helianthus bolanderi is a sunflower is a native to California and Oregon. It grows in two kinds of habitats. One found in the serpentine outcrops of the coastal ranges. These serpentine outcrops are soils with an extremely low calcium level and high content of magnesium and heavy metals. They sustain a highly specialized flora (Kruckberg, 1954). In addition *H. bolanderi* occurs as a ruderal weed in the central and northern valley of California at the same sites as *annuus*. In his studies (Heiser, 1949) has shown that these two species hybridize occasionally. Although the hybrid is largely sterile, it can produce some good pollen when *H. annuus* was introduced into California it hybridized with *H. bolanderi* which probably was then restricted to the serpentine outcrops. From repeated crosses and back crosses of these two species emerged a form of *H. annuus* that resembled *H. bolanderi* (Table 16.1, Fig 16.1) and adapted well to California environment, presumably by borrowing genes from *H. bolanderi*, from introgression of genes of *H. annuus* into *H. bolanderi* emerged the form of *H. bolanderi* that is capable of growing in disturbed habitats.

Table 16.1
Comparison of the number of Ray flowers and Disk Diameter of the common Sunflower Helianthus annuus, and the California Sunflower, H. bolanderi.
[Data from Heiser 1949.]

Species	Mean No. Ray flowers	Disk Diameter [cm.]
H. bolanderi- serpentine	12.7*	1.70*
H. bolanderi- valley	14.8*	2.07*
H. bolanderi x H. annuus hybrid	14-20**	2-3**
H. annuus, Calif.	18.2*	2.41*
H. annuus, non-Calif.	20.7*	2.98*

* *Mean value for garden grown plants (29 to 61 plants measured according to population).*

** *Range of values for 6 plants.*

In addition to *H. bolanderi* and *H. debilis* variety *cucumerifolius* the range of *Helianthus annuus* overlaps with other annual species. One of them is *H. petiolaris*, a species that is widespread in the western United States and occasionally eastward with distribution similar to *H. annuus* although this last species is more widespread. Although both species grow together in many places, there seem to be some slightly ecological preferences (Heiser, 1947).

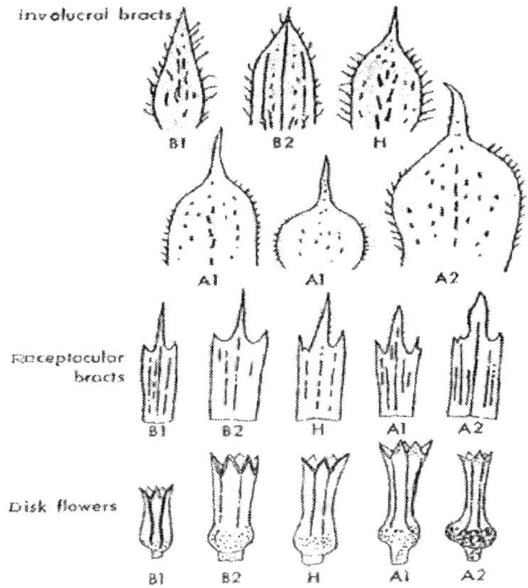

Fig. 16.1. Effect of hybridization on morphological characteristics of species, in this case *H. annuus* and *H. bolanderi.* B1, Helianthus bolanderi, serpentine race; B2, *H. bolanderi,* nonserpentine race, presumed to have introgressed with *H. annuus;* H, natural hybrid between *H. annuus* and *H. bolanderi;* A1, *H. annuus,* from California, presumed to have introgressed with *H. bolanderi;* A2, *H. annuus* from St. Louis, Mo., Outside the range of *H. bolanderi,* and consequently free of any contamination by *H. bolanderi.* Note how the presumed introgressed types look more like the hybrid than either of the "pure" species. (From Heiser 1949, by permission.)

In general *H. annuus* seems to be more restricted to heavy soils and *H. petiolaris* to sandy soils. In addition, while *H. petiolaris* comes into bloom in June and blooms through July and into August, *H. annuus* generally does not bloom until July and continues to bloom into September respectively somewhat the opportunities for gene exchange. Hybrid swarms between these two species are not frequent, but have been observed several times. Again the hybrid is quite sterile, but produces some viable pollen and consequently introgression has occurred. Introgressed populations of *H.*

petiolaris have broader and more serrated leaves, hispid involucral bracts, and a general increase in the size of all the parts of the plant. Populations of *H. annuus* subjected to introgression from *H. petiolaris*, on the other hand have narrower leaves, and a reduced number of ray flowers. (Table 16.2 Fig.16.2.)

Table 16.2
Comparison of the number of Ray flowers and Disk Diameter of the common Sunflower, *Helianthus annuus*, and H. *Petiolaris*.
[Data from Heiser, 1947.]

Species	Mean No. Ray flowers	Disk Diameter [cm.]
H. petiolaris	13-16	2.3
H. petiolaris x H. annuus	15-21	2.9
H. annuus	21-30	3.3

* *Range for several plants grown in the garden*
** *Mean for population from East St. Louis*

With these three examples of hybridization and introgression a pattern emerges that has been observed in several other species. The four species of sunflower, that have been discussed are most likely to have developed in isolation from each other under these circumstances they differentiated and became adapted to specific ecological conditions. However, they did not acquire total genetic isolation.

When *H. annuus* increased its range (probably as a result of man's activities) it came into contact with three other species. Hybrids were formed and they in turn backcrossed with the parents transmitting genes from one species to the other. Some of the genes transmitted confined on the recipient species characteristics that made it better adapted to the new environment. Because *H. annuus* was the invading species, it probably acquired physiological characteristics that allowed it to adapt better to the conditions of Texas, California and the Midwest. It in turn permitted these species to grow on disturbed soils (Fig 16.2) Harlan and De wet (1963) have termed species such as *H. annuus* compilospecies "robber species" because through hybridization they "borrow" genes from native species and in this way become more widely adapted.

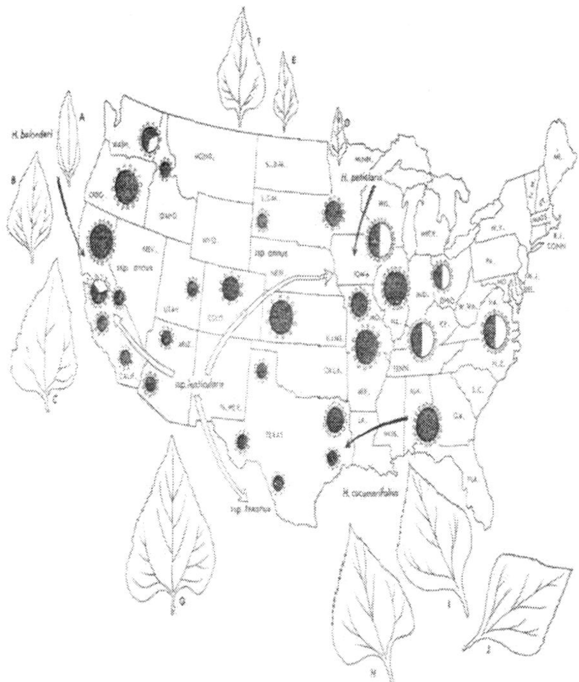

Fig 16.2 Probable evolution of *Helianthus annuus* in the United States according to Heiser. Subspecies *lenticularis* (or the derived subspecies *annuus*) spread from the southwestern United States to California where it hybridized with *H. bolanderi*. Subspecies *lenticularis* (or subsp. *annuus*) spread also to the midwest where it hybridized with *H. petiolaris*, and to Texas where it hybridized with *H. debilis* var. *cucumerifolius* giving rise to subspecies *texanus*. Differentiation as a result of selection was also taking place as can be seen by the different size heads, anther color, and number of ray flowers. Diagrams of heads drawn proportional to values of natural populations; "petals," each representing two ray flowers; shading represents anther color: black, purple anthers; hatching, red anthers; white, black anthers; combinations, mixtures of colors in the population. A–J: outline of leaves: A. *H. bolanderi*; B. *H. annuus* × *H. bolanderi*; C. *H. annuus* ssp. *annuus*; D. *H. petiolaris*; E. *H. petiolaris* × *H. annuus*; F. *H. annuus* ssp. *annuus*; G. *H. annuus* ssp. *lenticularis*; H. *H. annuus* ssp. *texanus*; I. *H. annuus* × *H. debilis* var. *cucumerifolius*; J. *H. debilis* var. *cucumerifolius*. (Data from Heiser 1947, 1949, 1951a, 1954.)

It can be seen then that the probable role of introgressive hybridization in plant evolution is to enable some species to acquire genes from other species and in this way increase their fitness and adaption to new habitat.

Selection of Breeding Barriers in Hybridization

How can the species that hybridize when they come into contact maintain their identity in the presence of gene flow, which tends to mix their genetic patrimony. The inability of the hybrid to grow in the habitat of either parent and its restriction to a belt of intermediate or disturbed conditions is one answer. The selection of mechanisms that will restrict hybridization still further is another possible way of protecting the identity of the parental species.

Helianthus annuus represents one outcome of hybridization the formation of a belt or zone of introgressed genotypes such as *H. annuus* variety *texanus*. Two species of sage of the genus Salvia in California represent another common situation, that of two species living side by side capable of interbreeding and forming vigorous hybrids. However, the hybrids are very few in actually, and there is apparently very little introgression.

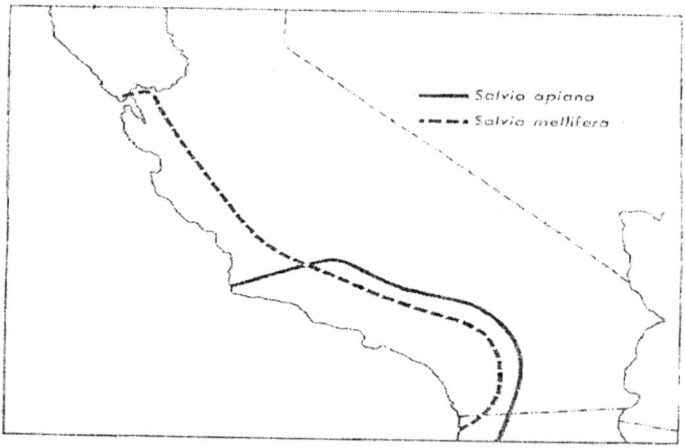

Fig. 16.3. Map of California showing the distribution of *Salvia mellifera* and *Salvia apiana* and the area of overlap of the two species (From Epling, 1947.)

Salvia apiana and *S. mellifera* are two species that grow in the coastal sage communities of Southern and Central California (fig16.3) their range overlap in the area between

Santa Barbara and the Mexican border. *Salvia apiana* shows a slight preference for the drier soils although the two species often grow intermixed (Epling, 1947, Andersorı, 1954)

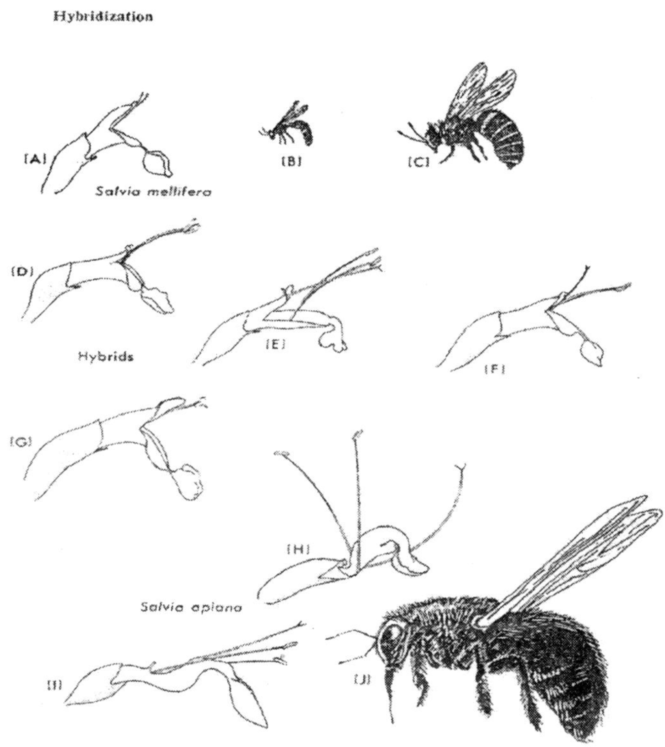

Fig. 16.4. Flowers of *Salvia mellifera, S. apiana,* and their hybrids and backcrosses, and some specific pollinators. **A.** *Salvia mellifera;* **B.** *Chloralictus* sp., A small-sized solitary bee that pollinates *S. mellifera;* **C.** *Anthrophora* sp., A medium-sized bee, pollinator of *S. mellifera;* **D.** F_1 hybrid between *S. mellifera* X *S. apiana;* **E-G.** Backcrosses of D. To both parents; **H-I.** *S. apiana,* untripped and tripped position of flower; **J.** *Xylocopa brasilianoram,* a pollinator of *S. apiana.* It can really be seen that the different size of flowers of these two species and their pollinators act as an isolating mechanism. (Redrawn from Epling, 1947a, and Grant and grant, 1964, by permission.)

Salvia millifera blooms earlier approximately from April to the end of May, where as *S. apiana* blooms from the middle of May to the end of June (Grant and Grant, 1961). Consequently there is little overlap in their blooming times. *Salvia mellifera* has a smaller flower whereas *S. apiana* has a much larger flower fig 16.4. Finally the first species is typically pollinated mainly by variety of small bees, flies and butterflies, *S. apiana* is pollinated mainly by larger carpenter bees. These ecological and mechanical differences combine to keep the two species reproductively isolated. Nevertheless, hybrids are formed occasionally most frequently in disturbed habitats. When these hybrids grow to maturity they are only partially fertile. Although in such instances some introgression occurs, it does not seem to be very widespread (Epling, 1947).

In case of two species of sage illustrate how species with and absolute genetic isolation barrier can grow side by side with little or no hybridization. This example also shows that there is no one absolute reproductive isolating barrier but a large number of partial barriers that reinforce each other to produce almost total reproductive isolation. Although few cases have been studied in such details as that of *Salvia*, many others are known. Many species of oaks are capable of interbreeding, so are many species of willows, poplars and birches. However, although hybrids are formed, the parental species maintain their identity and do not form homogenous hybrid swarms. As a matter of fact, it can be said that many species of closely related trees are interfertile but nevertheless do not hybridize extensively.

In annual plants the natural interspecific hybridization is rare. The probable reason for this has to do with still another fact of the phenomenon of hybridization; the effect that the fertility of the hybrid has on the total fitness of the population.

For the sake of survival, any population of plants has to replace those mature and reproducing individuals that die. It is well known that plants produce a great excess of seeds over the number that actually germinate and become mature plants. Consequently, under normal circumstances the production of a few or even a large number of more or less inviable seeds doesnot necessarily result in an immediate decrease in the number of the offspring. This is particularly true with long lived trees, because very few individuals have to be replaced every year. However annuals have to reproduce the population in its entirely every year. The production of a large number of hybrids could therefore be detrimental to an annual species, particularly if the backcrosses led to the formation of less fertile individuals. Because of annual plant species usually live in areas with large climatic changes from year to year, where certain seasons may allow the growth of only a small number of individuals, in the possible detrimental effects of hybridization is accentuated. This may explain why among annuals the species are usually well documented genetically even when they are not well demarcated morphologically. This is well illustrated by genera such as *Gilia* and *Clarkia* tables in which interspecific hybrids are often very sterile and, more important where natural hybridization seems to be rare or non existent.

How Reproductive Isolation develops in Hybridization

The hybridization is a very important phenomenon in plants. Basically it enables species to acquire new genetic material from outside sources when changes in selective pressure arise either from the migration of the species into new habitats or from long term changes in the environment. However, too high an incidence of hybridization could lead to the opposite effect a complete breakdown of the unique genetic patrimony of the species, making it less rather than more adapted to a particular environment. This is an occasional result of hybridization. More often the effect of hybridization is less drastic because of mechanisms that prevent indiscriminate mixing. These mechanisms are 1) an inability of the hybrid and the back crosses to compete successfully with the original species in all but a few special habitats. (such as disturbed sites or areas with environment intermediate between that of the two species and 2) a reduced fertility of the hybrid and of the backcrosses that slows down the mixing process. However, if the hybrid is successful in competing with the flora and becomes established and utilizes the resources of the environment, selection for higher fertility in the hybrid should occur.

In conclusion, hybridization can lead to the formation of new or modified species or it can lead indirectly to the formation of isolation barrier or it can proceed largely in a passive manner with slight effects, such as the formation of occasional hybrid swarms in areas of disturbed or intermediate habitats.

Hybrid Complexes

Groups of species in which hybridization has observed the basic diploid types have been termed collectively hybrid complexes by V. Grant (1953). He recognizes five types which are cited in order of taxonomic complexity.

Homogenic complex: in which the hybrid derivatives are sexually, entirely diploid

Colonal complex: in which the hybrids reproduce mainly entirely by vegetative division - localized as in *Opuntia.*

Heterogamic complex: where the products of hybridization are permanent structural heterozygotes as in *Oenothera.*

Polyploid complex (Bobcock and Stebbins, 1938): in which the derivatives of hybridization are sexual polyploids.

Agamic complex (Bobcock and Stebbins, 1938): unfertilized seeds in which the hybrids and their derivatives.

Taxonomic Treatment of Hybrids

ICBN of (1961) explicit about the methods that can be used to designate hybrids. The code states that hybrids between two species of the same genus are designated by formula and whenever it seems useful or necessary by a name thus putting the onus on the botanist to decide when a name is desirable. The formula consists of the names of the two parents connected by the multiplication sign (X) or the name of the genus followed by the two specific epithets connected by the same sign. When a name is used it must be validated by a Latin description, citation of type etc, in the same way as for any other binomial name. It is distinguished from a species by placing X before the second epithet e.g. *Salix X Capreola* = *Salix aurita* X *S. caprea* or *Salix aurita X caprea* (the latter is common practice). It is also a common practice of the generic name or the hybrid to combine euphoniously the name of the parent genera e.g. X *Festulolium loliaceum* (*Festuca pratensis* X *Lolium perenne*). Hybrids at generic level are named according to parentage not morphology. Whenever binary Latin names are used for hybrids, they cover all obvious manifestations of hybridity between the parents - i.e. F1 and later generation as well as backcrosses (e.g. *GeumX intermidium* covering the full range of hybridity between *Geum urbanum* and *G. rivale*). It therefore stands to reason that a binomial is more useful and biologically more meaningful for a more or less sterile F1 (Pauciform hybrid) than it is for a hybrid swarm (multiform hybrid).

17

Modern Theory of Evolution

Objectives

1. To understand and evaluate the misunderstood evolution
2. To understand analyse the synthetic theory of evolution
3. To follow the historical development of evolution.
4. To evaluate the structural evolution
5. To understand the concepts in terms of variation and evolution
6. To understand natural selection with scientific studies
7. To understand various principles, guidelines, basic assumptions of evolution
8. To analyze the evolution in terms of adaptation.

Convoluted Definition of Evolution

Higher organisms have evolved from lower organisms multicellular organisms have evolved from unicellular organisms.

The end products of this evolution is transformation of one organisms into another. Basically biological evolution is usually considered a series of processes involving desent of organisms with modification marked by competition and natural selection acting on variation. The general purpose of evolutionary study is to determine the origin of new taxa and to interpret the historical development of the old. The fundamental components of evolution are variation, speciation, and phylogeny with the basic activity of analysis of character variation.

Biologically, variation within a group of organisms is the divergence of structural, functional, and developmental characters from the typical mean or standard. Speciation is the development of populations of freely interbreeding organisms adapted to their environment that are reproductively isolated from other such populations. Phylogeny is the study of the genealogy and evolutionary history of groups of organisms.

Charles Darwin (1809-1882)

Charles Darwin published his monumental book, "on the Origin of Species by Natural Selection" in 1859 and it made the concept of organic evolution intellectually respectable. The development of the concept of evolution is interesting as a convoluted tale of people grouping at a persistent idea, but that is tangential to our story here. What is important is that the Darwin-Wallace theory of Evolution, as modernized by the incorporation of genetic theory, has basically stood the test of our time and has become virtual scientific dogma. Certainly the details are subject to revision, but the conceptual schemes of modern systematics rest firmly on the base of evolutionary theory. Biologists by and large accept the ideas that the nature of life has changed continually over the geologic ages and that life on earth consists of populations of organisms that are in various states of flux (Mayr, 1982).

Definition of Biological Evolution

A directional change in gene frequency over time in response to selective factors of the environment. Continuous genetic adaptation of the organisms to their environment.

The end products of modern theory of evolution is adaptation i.e. organic adaptation in varied environmental conditions to various selective factors.

The outcome of evolutionary process is adaptations of an organism to its environment. The evolution is the property of the population and net individuals. Natural selection is evolutionary force.

Definition of Modern Theory of Evolution

The modern synthetic theory of evolution has evolved during the last century through an accumulation of facts and theoretical conclusions from a number of scientist. Dobzhonsky (1937) in his book "Genetics and the Origin of Species" has emphasized the role of genetic changes in the population in the process of evolution. Julian Huxley (1924) and Ernst Mayr (1942-43) have explained the mechanism of origin of variations in high animals and Stebbins in higher plants. Cleveland Blackslee, Renner and others have been shown that a combination of gross chromosomal aberrations, rare combination, is balanced lethal systems and obligate self fertilization are important forces operating for the origin of species also operate for the evolution of genera, families and other higher categories.

Basic Process involved in Synthetic Theory of Evolution

At present the synthetic theory of evolution recognizes five basic process namely 1) gene mutations 2) changes in chromosome number 3) genetic recombination 4) natural selection and 5) reproductive isolation.

The three accessory processes also effect the marking the process. These are migration, hybridization and chance in small population.

Othneil C Marsh (1831-1899), Cope (1840-1897), Mathew Gregory, Romen and Simpson in America, Wood ward (1864 - 1944) and Watson in England and Broom (1866-1951) in South Africa has contributed immensely to vertebrate paleontology. J.B.S Haldane, Fisher and Sewal Wright and S.S Chetiverikov have provided mathematical evidences to emphasize the role of natural selection in the evolution of population. Studies on genus *Crepis* by E.B. Bobcock.

In a given generation the genes transmitted by those individuals who most successively reproduce, will predominate. It means genetic composition of a population becomes altered each successive generation. It has changed in past, is changing today and will continue to change in future. This ever changing phenomenon is evolution. It can be defined as the changes in the genetic composition of a population with the passages of each generation. Evolution through the genes is the way through which stable adaptation comes about. The end products of evolution is organic adaptation.

Structural Evolution

Structural evolution refers to the historical sequence of changes in morphology of organs, tissues or cells – adaptive significance of spines, prickles, thorns, needles, arms, multilayered epidermis, palisade parenchyma, transfusion tissue, scattered vascular bundles, intraxylary phloem, intruded phloem, sunken stomata, epidermal indumentum, intrafascicular cambium and sclereids, idiobasts, such structures like, phyllode, cladode, phylloclade such structures are useful for better adaptations were brought into the organisms via genes. Structural evolution may be deduced from fossil evidence and from analysis of characters and character states of extent taxa, including the stages in a development of a structure. A consideration of structural changes that have occurred during evolutionary history adds insight and meaning to the taxonomic categorization of plants. The study of structural modifications may be important understanding the possible adaptive significance of such changes in the evolution of organisms within a particular environment and with reference to interactions with other organisms.

The evolution of structure is always studied with regard to group of organisms and therefore can be determined by using the methodology of phylogenetic systems. Within the vascular plants, heterospory and secondary growth are considered to be derived structural features although both apparently evolved independently in different lineage.

Variation and Evolution

Biologically, variation within a group is the divergence from the standard, mean or typical structural, functional and developmental characters particularly those not attributable to differences in age, life history, sex, or environment. Systematists ascertain types, patterns and trends in variation of selected characters within a populations, species and higher groups to delimit, circumscribe, and classify taxa as well as to determine the phenetic and phylogenetic relationships.

Evidentiary characters used in variation studies may be classified as physical (shape, color) chemical (process, composition) and biological (structure, function) character variation studies, evidence is information is used for systematic purposes. Phenotypic variation is morphological functional or developmental variation resulting from the action of different environments on one or more genotypes within populations.

b) Genotypic variation (genetic) is the differences in genotypes within a population(s) as a result of mutation and recombination.

Sources of genetic variation in natural populations are due to mutation and recombination. Mutation is the occurrence of heritable change in the genotype of an organism that was not inherited from its ancestors. In theoretical sense, mutation is the ultimate source of genetic variability. All genetic changes in existing individuals and populations except those due to recombination or mutation. A mutation may be as minute as the substitution of a single nucleotide pair in the DNA molecule or as great as a major change in chromosome structure or number (gene and chromosomal mutations respectively) chromosomal mutation may be due to deletion, inversion, aneuploidy or polyploidy.

Recombination is a reassortment of chromosomes, crossing over segments and therefore genes that produce new genotypes in sexually reproducing organisms. Recombination is immediate source of variation observed in populations (such as height of plant and color of corolla) and is the raw material on which the process of evolution and speciation act. Factors controlling recombination in plants (Grant, 1981) are operative at meiosis (chromosome number, frequency of crossing over, hybrid sterility) and fertilization (breeding system, pollination system, dispersal potential, population).

Recombination also results from hybridization and introgression. Natural hybridization provides gene flow between individuals from divergent populations at the local race to taxonomic intraspecific and specific levels (and occasionally at the genetic level in plants). Introgression is a process of successive hybridizations that cause the migration of genetic material from one species into another. Initial interspecific hybridization usually followed by a series of backcrosses to one of the parents (or it can occur independently in both directions).

Mutations and genetic recombination produce the genotypic variation in populations basic to the onset of the speciational process. Natural hybridization furnishes the genetic material from established taxa for the possible development of new species under different selective pressure.

Mutation never seems to happen in response to a need for particular genetic change, nor is particular mutation released by a particular constellation of environmental factors. This is still a great deal to learn about the process of mutation. Yet there is little reason to doubt that most mutations are merely copying mistakes during the replication of the genetic material. Hence it connects to state that mutations are strictly accidental occurrences.

Variations in Chromosomes and Genes

In sexually reproducing higher organisms these genes are organized in chromosomes. The re-assortment genes within the chromosomes, which normally happens once in every generation, is governed largely by chance. Each individual in sexually reproducing organisms has two homologous sets of chromosomes, one from the father and one from the mother. At some times prior to the formation of gametes, two homologous chromosomes exchange equivalent pieces with each other by process called "crossing over". By and large no laws seem to determine where the chromosome will break or how large the pieces will be that are exchanged which particular combination of pieces of maternal and paternal chromosomes making up the new chromosomes will enter a given egg or spermatozoan is largely a matter of chance atleast in most of chromosomes and most species. Likewise, it is largely a matter of accident which chromosome will go into which germ cell, provided only that each cell receives its full set of chromosomes.

Numerous mechanisms evolved in due time to make recombination increasingly precise in every aspect of diploidy through two homologous chromosome sets, one derived from the father, the other from the mother, of an elaborate process of meiosis during which homologous chromosomes exchange pieces, so that the chromosomes of father and mother are transmitted with a novel assortment genes. These mechanisms regulate genetic recombination among individuals by far the major source of genotypic variability in higher organisms. The enormous power of the process of genetic recombination of sexual reproduction becomes evident if we remember that in sexually reproducing species no two individuals are genetically identical.

Natural Selection and Other Selections

"The preservation of favourable individual differences and variations and the destruction of those which are injurious - I have called this as Natural Selection." - Darwin

Natural selection

1) **Stabilizing selection:** The phenotypes with characteristics near the mean of the population are more fit than those at the extreme an account of a stable environment. As a result of this kind of selection, the mean of the population will not vary although its variability may, depending on the genetic determination of the character under consideration.

2) **Disruptive selection:** In an environment favoring the extremes, any mechanisms that would keep opposite extremes from breeding with each other would be highly advantageous leading to the second type of selection is called disruptive selection.

3) **Directional selection:** This is the type of selection where one extreme phenotypic expression of a character is favoured, which happens in nature when the environment is changing constantly in one direction. Directional selection produces phenotypic and genotypic changes more rapidly than any other kind of selection.

Other selections

From the view of genetics, evolution is the change in the frequency of the genes of a population. In order for change to occur there must be at sometime a source of new genes. The sources are two 1) Mutation and 2) Introduction of genes from neighboring population in the form of seeds or pollen (Fig 16.2). Further more in order to have change, there must be a force producing change, the force is off course natural selection, although occasionally change solely as a result of chance phenomenon known as genetic drift.

Genes do not occur by themselves. They are integral part of the chromosome of the plant. Likewise natural selection does not act directly on the genes, but it affects phenotypes (Fig 16.2). Through the process of sexual reproduction recombination, the genes of population are shuffled and reshuffled in each generation into novel combination of genotypes that give rise to new kinds of phenotypes.

Fundamental Principles of Evolution

1) Existing species are products of evolutionary forces that have descended from the pre existing species
2) Variation in organisms makes the evolution possible
3) Sources of variation in natural population are genetic and environmental
4) Limitation of variation in natural population are intrinsic and due to such causes as self fertilization, asexual reproduction, genetic fixation and stabilizing selection.

Guidelines of Evolutionary Study

1) Evolution may tend toward elaboration and diversity or toward reduction and simplicity of organisms or organismal plants.
2) A character may evolve at different rates in different taxa and within one taxon various characters may evolve at very different rates and in different directions.
3) Evolutionary trends, progressive or retrogressive are

generally consistent but occasionally reversible with changes in the environment.

4) All parts of organisms at all stages of their development may produce evidence of evolutionary significance.

Basic Assumptions for Evolutionary Study

Based on Charles Darwin, on the Origin of Species, 1859

1) Many more individuals are born in each generation than will survive and reproduce.
2) Variations occurs among individuals, they are not identical in all of their characteristics.
3) Some individuals with certain characteristics have a better chance of surviving than individuals with other characteristics.
4) Some of the characteristics of individuals resulting in differential reproduction are heritable.
5) Extremely long spans of time are available for slow, gradual change in individuals.

Organic Adaptation followed by genetic adaptation in population and organisms is called adaptation.

Adaptive Characters in terms of Advanced Nature

Primitive Characters	Advanced Characters
1. Woody plants- large trees	Herbaceous
2. Sparingly branched	Numerous slender branches
3. Leaves simple, Pinnately nerved	Compound leaves, Palmately or Pinnately compound
4. Net, Conconduplicate venation lamina folded	Parallel venation
5. Node: Tri or penta locunar solitary flowers	Unilocular type of node
6. Cymose branching - simple cymose	Racemose type - Compound cymose
7. Spiral arrangement	Cyclic arrangement

8. Actinomorphic	Zygomorphic
9. Vascular bundles in a single ring	Scattered vascular bundles
10. Petals and numerous spirally arranged laminar stamens	Petals are fused or reduced
11. Uniaperturate type of pollen	Triaperturate type of pollen
12. Pollinated by beetles	Wind pollinated
13. Two integuments in ovule	Uni-integumentic ovule
14. Paracytic stomata	Polygonum type of embryo sac
15. Cyclic vascular bundles in a ring	Scattered vascular bundles
16. Uni aperturate many carpels	Triaperturate
17. Axile placentation	Reduction in number of carpels
18. Copious endosperm	Ruminate less endosperm
19. Dehiscent fruits	Indehiscent fruit
20. Unisexual flowers	Bisexual flowers
21. Apocarpy	Syncarpy

Thorne's Traits

Thorne (1958)

1) Existing species have descended with change from pre existing species and are the products of evolutionary force.
2) Ancestral conditions and trends of specialization are recognizable in the organs, tissues and cells of living and fossil angiosperms.
3) The presence of vestigial rudiments of organs or sometimes the presence of vestigial vascular supply to greatly modified or missing organs or loss or fusion or other.
4) All parts of plants at all stages of their development may produce evidence that is valuable in establishing relationships.
5) Evolution may tend toward elaboration and diversity or towards reduction, and simplicity.
6) The rate and direction of evolution vary in the different

organs and tissues of plants

7) Most existing angiosperms highly specialized than their primitive ancestors
8) Evolutionary trends are sometimes reversible under the influence of changes in environmental factors
9) Once lost, organs usually are not regained
10) New angiosperm structures have arisen as modification of or as out growths from pre-existing structures
11) The sporadic or restricted occurrence of unusual or uncommon characteristics lacking apparent evolutionary significance
12) The occasional attainment of certain levels of evolutionary development is frequently valuable in determining the affinities of families and orders
13) Embryos and seedlings of related though dissimilar plants often resemble each other more than do the adult plants because of their apparent retention of primitive characteristics.

Smith's Traits Suggested as Adaptive Characters

- Temperate plants
- Climbing or herbaceous
- Woody heteroxylous (Vesseled)
- Binneal or annual habit
- Aquatic, epiphytic or Saprophytic, parasitic habit
- Vascular bundles scattered
- Chlorophyll absent
- Node variously altered
- Leaves opposite whorled
- Leaves compound
- Flowers unisexuals
- Flowers inflorescences
- Flowers anemophilous

- Floral parts whorled or valvate
- Flowers oligomerous
- Perianth differentiated into calyx and corolla or further reduced
- Flowers without petals
- Petals connate
- Zygomorphy
- Perigyny, epigyny
- Stamens few
- Stamens connate
- Stamens variously modified
- Pollen tricolpate
- Carpels few
- Carpels connate
- Carpels variously modified
- Axile or parietal placentation
- Tenuinucellate ovule
- Shorter sieve tube elements with specialized sieve plates
- Higher chromosomal number
- Fruits aggregate
- Fruit capsule, berry, drupe
- Seeds small
- Embryo well developed
- Endosperm little no cellular endosperm
- Helobial endosperm

The evolutionary trends that are given as traits from various classification systems of Hutchinson, Takhtajan, Cronquist and Dahlgreen can be visualized.

GLOSSARY

adaptation – the condition of showing fitness for a particular environment, as applied to characteristics of a structure, function or entire organism, also the process by which such fitness is acquired.

agamic – a species or generation which is not reproducing sexually.

agamospecies – a species without sexual reproduction, an asexual species.

alleles – one of a pair or series of forms of genes that are alternative in inheritance because they are situated at the same locus in homologous chromosomes.

allopatric – said of taxa whose geographical ranges do not overlap.

allopatric speciation – species formation during geographical isolation (eg. sympatric speciation)

allopolyploidy – a type of polyploidy characterized by containing genetically different sets of chromosomes from different species.

alpha taxonomy – the level of taxonomy concerned with characterization and naming of species.

amino group – NH_2 a chemical group characteristically basic because of the addition of portion to form ammonium ($-NH_3^+$)

amphiploid – a polyploidy produced by the chromosome doubling of a species hybrid, that is of an individual with two rather different chromosome sets.

anatomy (plant) – the study of the internal structure of a plant.

aneuploidy – the existence of more (or less) than the basic chromosome number in a normal plant, which is not an exact multiple of the basic number.

antigen – a substance capable of inducing the formation of antibodies when introduced into a blood stream of an animal.

anthocyanin – a water soluble pigment, characteristic of red and blue coloured flowers.

apomixis – reproduction in which sexual organs or related structures take part but fertilization does not occur, so that the resulting seed is vegetatively propagated.

appressed (pressed down) – closely and flatly appressed.

autogamous – self fertilizing.

autosome – one of the chromosomes other than a sex chromosome.

autopolyploidy – a polyploidy arising through multiplication of the complex haploid set of a species.

betalains – a kind of nitrogenous pigment found in the order-Centrospermae it replaces the anthocyanins in that group.

beta taxonomy – the level of taxonomy concerned with the arranging of species into a natural system of lower and higher taxa (of alpha taxonomy, gamma taxonomy).

biological classification – the arranging of organisms into taxa on the basis of inferences concerning their genetic relationship.

biosystematics – the application of genetics, cytology, statistics, and chemistry to the solution of systematic questions in order to provide explanations about the diversity of organisms within the frame of the theory of evolution.

biological species concept – a concept of the species category stressing reproductive isolation and the procession of a genetic program effecting such isolation.

biosystematy – see biosystematics.

breeding population – a reproductive community of sexual and cross fertilized individuals that share in a common gene pool.

carbohydrates – organic compounds with the empherical

formula $(CH_2O)n$. Examples of carbohydrates : sugars, starch, cellulose.

carboxyl group – R-COOH. a chemical group characteristically acid, as a result of the dissociation of the hydroxyl H to form R-COO.

carotenoid – a large group of compounds that constitute that yellow to red fat–soluble pigments of plants.

carpel – one of the foliar ovule bearing unit of a pistil or ovary.

category – any of the levels in the taxonomic hierarchy, such as species.

centromere – the constriction in the chromosomes that is involved in the movement of the chromosomes to the poles during cell division.

character – an attribute of an organism resulting from the interaction of a gene or genes with the environment.

—continuous – set of a character whose variation in the population does not show discontinuous.

—discontinuous – said of a character whose variation in the population can be broken down into discrete classes.

- qualititative – see discontinuous character.
- quanititative – see continuous character.
- state – any of the values that a particular character can have. Used particularly when referring to discontinuous characters.

chiasma – an exchange of partners between paired chromatids in the first division of meiosis.

chromatid – one of the threadlike structures formed by the longitudinal division of a chromosome during meiotic prophase and known as daughter chromosomes during anaphase.

chromatin – general term used to refer to the chromosome substances the stain with such dyes as acetocarmine or acetoorcein. The nucleoprotein complex of the chromosomes.

chromosomes – structural units of the nucleus that carry the genes in linear order.

chromosomal inversion – reversal of the linear order of the genes in a segment of a chromosome.

cladism – a taxonomic theory by which organisms are ordered and revised actively on the basis of 'recency of common descent' that is on the basis of the most recent branching point of the inferred phylogeny.

cladogram – a dendrogram based on the principles of cladism. A strictly genealogical cladogram in which rates of evolutionary divergence are ignored.

classification – the ordering of plants (or any other objects) into classes or groups.

clone – a group of individuals resulting from vegetative multiplication of any plant propagated vegetatively and therefore presumably a duplicate of its parents.

clusting method- methods of grouping related or similar species into species, groups or higher taxa.

continuous variation – variation in which individuals differ from each other by infinitely small steps; as variation in quality of expression of a character or group of characters.

crossing over – the exchange of corresponding segments between chromatids of homologous chromosomes during meiotic prophase. Its genetic consequence is the recombination of linked genes.

cytokinensis – the process of cell division as contrasted with the division of the chromosomes and the nucleus.

cytogenetics – the comparative study of chromosomal mechanisms and behaviour in populations and taxa and their effect on inheritance and evolution.

cytology – the study of the cell.

Darwinian principles – the taxonomic importance of characters that are not the result of a specific adhoc adaptation.

data matrix – a tabulation of differences between species (or other taxa in rows (character) and columns taxa.

delta – symbol; usually used to denote very small change.

deme – a local population of species; the community of potentially interbreeding individuals at a given locality.

denaturing (proteins) – the loss of the native configuration of macro-molecule resulting for instance, from heat treatment, extreme PH changes chemical treatment and so on. It is usually accompanied by loss of biological activity.

dendrogram – a diagrammatic drawing in the form of a tree designated to indicate degrees of relationship (phylogenetic).

diakinesis – the last stage of the prophase of meiosis. During diakinesis chromosomes usually can be observed very well under the microscope.

decline – unisexual requiring two flowers to represent both sexes.

derived character – a character that differs materially from other ancestral condition.

dichotomous – divided or dividing into two parts.

dimorphism – occurance of two distinct morphological types. (morphs, phena) in a small population (if sexually dimorphic) polymorphism.

dioecious – plants in which staminate (male) and pistillate (female) flowers occur on different individuals.

diploid – the chromosome state in which each type of chromosomes except for the sex chromosomes is always represented twice.

disk flower (in Asteraceae) – the tubular flower in the center of the flowering heads of most members of the sunflower family (compositae).

DNA (Deoxyribonucleic acid) – a polymer of deoxyribonucleotides. The genetic material of all cells and most viruses.

dominant (gene) – intra allelic interaction causing one allele to manifest itself more or less, when heterozygous, than its alternate allele.

ecological isolation – a condition in which interbreeding between two or more otherwise sympatric population is believed to be prevented by mating in different ecological niches (if reproductive isolation, geographic isolation).

ecological race – see ecotype.

ecophenotypic variation (habitat variation) – a non genetic modification of the phenotype by specific ecological conditions, particularly those of habitat.

ecospecies – a group of populations so related that they are able to exchange genes freely without loss of fertility or vigour in the offspring (Turesson).

ecotype – a group of populations that share certain physiological and genetic characteristics that make them better adapted to the ecological conditions of an area than other populations of the species.

edaphic factor – the influence of soil properties or organisms (especially plants).

embryosac mother cell – the cell that through meiosis and subsequent mitosis give rise to the embryosac.

endopolyploidy – increase in the chromosome number of vegetative cells.

environment – the sum total of external conditions that effect growth and development of an organism.

epistasis – dominance of one gene over a non allelic gene. The gene suppressed is said to be *hypostatic*. More generally the term *epistasis* is used to describe all types of interallelic interaction whereby manifestation at any locus is affected by genetic phase at any or all other loci.

form – a natural form for a single individual, phenon or taxon (neutral term).

family (nomenclature – a required category formed by one or more genera.

(genetics) - a group of individuals directly related by descent from a common ancestor.

fitness – the ability to reproduce fertile offspring that survive to reproductive age. Also called *Darwinian fitness*.

flexibility, long range – the ability of maintaining.

founder principle (Mayr's) – the phenomenon by which the gene pool of a newly established population represents only a small portion of the species gene pool.

gamma taxonomy - the level of taxonomy dealing with various biological aspects of taxa ranging from the study of interspecific populations to studies of speciation and evolutionary rates and trends.

Gause's rule – the theory that no two species with identical ecological requirements can co-exist in the same place (competitive exclusion).

gametophytic generation – the generation in plants that bears the sex organs; in ferns a small body bearing archegonia and antheridia in the angiosperms the pollen and pollen tube and the embryosac.

gene flow – the transmission of genes from one population to another.

gene frequency – the percentage of a given gene in a population (gene, population).

gene pool – the locality of the genes of a given population existing at a given time.

genetic background – the genotype with which a particular gene interacts.

genetic death – refers to the loss of gametes when one allele is replaced by another in a population.

genetic drift – the phenomenon by which each generation is a random sample of the previous one.

genetic load – the amount by which the average fitness in the population is lower than that of an individual with the optimal genetic composition.

genetic locus – the position occupied by a gene in a chromosome.

genome – a set of chromosomes corresponding to the haploid set of a species.

genotype – the entire genetic constitution of an organism as distinguished from its physical appearance or phenotype.

genus (nomenclature) – a required category formed by one or more species.

geographic isolation – the separation of a gene pool by

geographic barriers, the prevention of gene exchange between population and others by geographic barriers.

glabrous – not hairy.

gynodioecious – a plant that has pistillate (female) flowers and a few bisexual flowers.

haploid – the chromosome state in which each chromosome is present only once.

heterochromatin – the dark staining areas of a chromosome. The chemical and physical properties of heterochromatin are still very much debated.

heterokaryon – a mycelium of a fungus with two or more genetically different nuclei within each cell.

heteromorphic – with more than one form.

heterosis – hybrid vigour that causes an F_1 hybrid to fall outside the range of the parents with respect to some character or characters usually applied to size, rate of growth, or general fitness, although in strict terms it should be applied only to fitness, and the term *luxuriance* used to denote size or rate of growth.

heterozygous – having unlike alleles at one or more corresponding loci (opposite of homozygous).

homologous (chromosomes) – chromosomes that have homologous segments and may pair at meiosis, but are homologous along their entire length. Usually belonging to different species.

homogametic- producing only one kind of gamete.

homozygous – having like alleles at corresponding loci on homologous chromosomes. An organism can be homozygous at one or many loci.

hybrid superiority – see heterosis.

hypostasis (hypostatic) – recessiveness of a gene with respect to a non allelic gene. The converse of epistasis.

inbreeding – the mating of individuals more closely related than individual mating at random. The mating of a plant with itself.

incompatibility alleles – the series of genes, usually referred to as S alleles, that prevent the pollen of certain plants from germinating on the style of the same plant that produced the pollen.

individual variation – variation within a population.

internodes – the part of a stem between two nodes, or joints, where leaves are borne.

intraspecific – within the species, usually applied to categories (subspecies) and phenon (varieties).

introgressive hybridization – the introduction into one species, through backcrosses, following upon an interspecific cross, of genes of another species.

inversion (Chromosomes) – a rearrangement of a chromosome segment so that its genes are in reversed linear order.

ionic bond – the electrostatic forces acting between oppositely charged ionic chemical groups.

isolating mechanism – any of many structures or phenomenon that separate one plant (or population) from another so that mating between them is prevented.

isophene – a line connecting points of equal expression of a character lines at right angles to a line on a map.

karyology – the study of the cell nucleus.

karyotype – the chromosomal complex characteristic of a group of allied plants, associated with morphology and number of chromosomes.

linkage – the phenomenon where genes are located on the same chromosome and therefore tend to be transmitted together in the gametes.

mean – the mean is the average, which is found by adding the series of numbers, and dividing the total by the number of values. The mean of leaf length/ width ratio.

meiosis – the process by which a diploid cell gives rise through two consecutive divisions to four haploid spores.

meristem – undifferentiated tissue whose cells are capable of developing into various organs or tissues.

metaphase – the second stage of cell division, in which chromosomes align in the middle of the cell plate prior to separation (or division) and movement to the poles.

microsporocytes – the cells that give rise by meiotic division to the smaller of two kinds of spores. In the angiosperms, the cells that give rise by meiosis to the pollen.

mitosis – the process by which the nucleus of a somatic cell divides into two nuclei with the same number of chromosomes as the mother cell. Mitosis is usually followed by cytokinesis.

modifier (genes) – genes that effect the expression of a non allelic gene or genes.

monoecious – of the population or species that has plants with staminate (male) flowers and pistillate (female) flowers, on the same plants.

monoecious decline – of the population or species that has the staminate (male) flowers and pistillate flowers borne on the same plant but on separate flowers.

monograph – in taxonomy, an exhaustive treatment of a higher taxon in terms of available information pertinent to taxonomic interpretation; usually involving full systematic treatment of the comparative anatomy, biology, ecology, and detailed distributional analysis of all included taxa.

monomorphic – having only one form.

morphology – the study of the form of plants.

multiple factors – see multiple gene system.

multiple gene systems – systems of gene whose effects are too slight to be identified individually but that through similar and supplementary effects can have important effects on total variability.

multivariate analysis – the simultaneous analysis of several variables characters.

mutation – a sudden inheritable variation in a gene or in chromosome structure. The product is called a *mutant gene.*

back mutation – a mutation of a mutant gene to its original state.

recurrent – a mutation that occurs regularly in a population.

mutant genotype – a genotype in which mutations occur at a very high frequency.

natural selection – preservation of favourable individual differences and variations and the destruction of those which are injurious.

nomenclature – the application of distinct names to each of the groups recognized in a classification.

oligogenic character – a character determined by only few genes.

ontogeny – the development cycle of an individual.

outbreeding – the habitual crossing with another plant.

pachytene – the third stage of meiotic prophase, when homologous chromosomes pair and crossing over takes place.

papilionate flower – the pea like (butterfly like) flower of the tribe Papilionoideae of the pea family. (Leguminosae).

pappus (Asteraceae) - peculiar modified calyx borne on the upper part of the ovary and persisting in the fruit, it can be plumose bristle like and so on, and aids in seed dispersal.

peptide bond – a covalent bond between two amino acids in which the alpha amino group of one amino acid is bonded to the alpha carboxyl group of the other with the elimination of water.

petiole – the leaf stalk, missing in some leaves.

phenetic – relation of similarity of characters, as opposed to phylogenetic or similarity of descent.

Phenogram – a diagram indicating degree of similarity among taxa.

Phenetic ranking – ranking into categories, strictly based on degree of overall similarity.

phenotype – appearance of an organism, the result of the development of the genetic message contained in the genotype in a particular environment.

phylogeny – the evolutionary history of a population, species or higher category. Also used when referring to the evolutionary history of a structure or organ.

physiology (plant) – the study of the function of plants.

pin – the long styled form in a distylous, heterostylous form.

pleiotropy – the effect of one gene on more than one phenotypic character.

polygamodioecious – a plant that has unisexual flowers and a few bisexual flowers.

polymer – a regular, covalently bonded arrangement of basic sub units flowers called monomers, which is produced by the repetitive action of one or a few chemical reactions.

polypeptide – a polymer of aminoacids linked together by peptide bonds.

polyploid – a plant with a chromosome complement of three or more sets of chromosomes each with the exact haploid number.

population – see breeding population.

protandry – maturing of the anthers before the style, which usually impedes self fertilization.

protogyny – maturing the styles before the anthers which usually impedes self fertilization.

pubescent – covered with short soft hairs, downy.

race – subspecies.

ray flower – outer flower with a strap like corolla in the flowering head of some members of the sun flower family (Compositae).

recessive (gene) – the member of an allelic pair that is not expressed when the other (dominant) member occupies the homologous chromosome.

recombination – formation of new combination of genes as a result of the sexual process; the appearance of new combinations of traits in the offspring also applied sometimes to the process of hybridization between members of the same population and species.

regressive analysis – a form of multivariate analysis.

regressive character – a character which is being reduced or lost in the course of phylogeny some times independently in several related times.

relationship – a statement about two or more objects that is either true or false.

reproductive isolation – a condition in which interbreeding between two or more populations is prevented by intrinsic factors.

reticulate evolution – evolution dependent on repeated intercrossing between a number of lines, and thus both convergent and divergent at once.

ribonuclease – the enzyme that acts on RNA (ribose nucleic acid) as substrate.

scatter diagram – a bivariate or multivariate graphic method of population analysis.

self comptability – capability of producing seeds when the style is dusted with pollen from the same plant.

self incompatibility – genetic incapacity of producing seed when the style is dusted with pollen from the same plant.

semispecies – the component species of subspecies (Mayr); also, populations that have acquired some, but not yet all, attributes of species rank, border line cases between species and sub species.

serology -- the study of nature and interactions of antigens and antibodies.

sessile – not stalked sitting.

shift (chromosome) – the shifting of chromosome segment to another position with same chromosome.

sibling species – species that are morphologically identical or nearly identical in their phenotypes.

species – the basic required unit of the taxonomic system, below the rank of genus.

speciation – the process of multiplication of species, the origin of discontinuities between populations caused by the development of reproductive isolating mechanisms.

standard deviation (SD) – the square root of the sum "of the squared deviations from the mean, divided by N. SD = "Ed^2/ N.

standard error (of the mean) – standard deviation divided by the square root of the sample size N. SE = SD / " N.

sporophytic generation – the generation that produces spores. In fern and seed plants the foliaceous vegetative plant as opposed to the gametophyte.

subspecies – a monophyletic group of entirely or largely allopatric species (eg: semispecies allopatric).

supraspecific – a term applied to a category or evolutionary phenomenon above the species level.

sympatric – set of taxa whose geographical ranges overlap.

systematics – the branch of biology concerned with the comparative study of organisms and all relations among themselves.

taxon – a general term applied to the organisms in any taxonomic element. Population, species, genus, and so on, irrespective of the category in which it is placed.

taxonomy – the study of classification, its principles, procedures, rules.

tetraploid – a polyploidy with exactly four sets of basic (haploid) chromosomes.

triploid – a cell or individual with three haploid sets, one of the forms of polyploidy.

thrums – the short styled form, in a distylous heterostylous species.

translocation – change in position of a chromosome segment to a position in a different chromosome.

> *non reciprocal* – when only one segment is exchanged between two chromosomes.
>
> *reciprocal* – when both the donor and the recipient chromosome exchange segments.

trinucleate – having three nuclei.

trophic – pertaining to a relation of dependence, such as herbivorous animals dependence on plants, carnivores dependence on herbivores.

univalent (chromosome) – a chromosome that during meiotic pachytene does not pair.

variance – the square of the standard deviation.

variety – an ambiguous term of classical taxonomy for a heterogenous group of phenomenon including non genetic variations of the phenotype, morphs, domestic breeds, and geographic races.

zygote – first cell of an organism, formed by the union of the gametes from which in multicellular organism the whole individual will develop.

BIBLIOGRAPHY

Adanson, M. 1847. Families des plantel 2 vol. Paris.

Ahmad, K. J. 1974. Cuticular studies in some species of *Hemigraphis* and *Strobiolanthes. Proc. Indian Acad. Sci. (Sect. B).* **79**:29-40.

Alston, R. 1965. A symposium on macromolecular evolution: a systematic perspective. *Biosciences* **15**:466-468.

Alston, R. and R. N. Lester, 1963. Natural hybridization among four species of *Baptisia Amer. Jour. Bot.* **50**:159-173.

Alston, R. E. 1966. Chemosystematics or biochemical systematics in comparative Tsuan ed, *pp. 33. 56p. Academic press.* London phytochemistry .

Alston, R. E and B. L. Turner. 1963a. **Biochemical systematics**. Prentice Hall. Engelwood Cliffs, N. J.

Alston, R. E and B. L. Turner. 1963b. Natural hybridization among four species of Baptisia (Leguminosae) *Amer. J. Bot.* **50**:159-73.

Amadon, D. & L. L. Short. 1976. Treatment of subspecies approaching species status. *Syst. Zool.* **25**:161-167.

Amadon, D. 1943. **Specialization and evolution**. *Amer. Nat.* **77**:133-141.

Anderson, D. E. 1949. Introgressive Hybridization. New York.

Anderson, D. E. 1952. **Plants, Man and Life**. Boston.

Anderson, D. E. 1953. Introgressive Hybridization. *Biological Reviews* **28** : 280-307.

Anderson, D. E. and Anderson, B. 1954. Introgression of *Salvia apiana* and *Salvia mellifera. Ann. Mo. Bot. Gard.* **41** : 329 – 338.

Andrea Cesalpino; 1583. **De plantis**.

Aristotle.

Avdulov, N. P. 1931. Karyotype-systmatsche Untersuchunger der

Families. Oramineen. *Bull. Appl. Bot. Genet. Pl. Breed. Suppl,* **43** : 1-428.

Ayensu, E. S. 1970. Comparative anatomy of *Dioscorea rotundata* and *Dioscorea cayenensis.* In N. K. B. Robson et al (eds). New Research in Plant Anatomy *pp.* 127-136. Linnean Society, London.

Azizian, D and D. F. Cutler. 1982. Anatomical, cytological and phytochemical studies on *Phloemis* L. and *Enemostachys* Berge (Labiatae) *Bot. J. Linn. Soc.* **85.** 249-281.

Babcock, E. P. etal. 1947. The Genus *Crepis.* Part one. The taxonomy, phylogeny, and Evolution of *Crepis. Uni. Calif. Publ. Bot.* 21, 67, 68, 99, 101, 199, 204.

Baker, H. G. 1948. Dimorphism and monomorphism in the plumbaginaceae I : a survey of the family. *Ann. Bot. H.* **12** : 207-219.

Baker, H. G. 1953. Dimorphism and monomorphism in Plumbaginaceae III: a survey of the family. *Ann. Bot. II.* **17** : 645-627.

Baker, H. G. 1955. Self compatibility and establishment after long distance dispersal *Evolution.* **9** : 347-348.

Baker, H. G. 1959. Reproductive methods as factors in speciation in flowering plants. *Spring Harb. Sym. Quant. Biol.* **24** : 177-191.

Baker, H. G. 1961. Rapid speciation in relation to changes in the breeding system plants. Toronto Uni. Press. **Recent Advances in Bot.** 881-885.

Baldwin, J. M. 1902. **Development and Evolution**. Macmillan, London.

Baldwin, J. T. 1938. *Kalanchoe*: The genus and its chromomes. *Amer. J. Bot.* **25**:572-579.

Bannes, B. V. 1961. Hybrid aspens in the lower peninsular of Michigan- Rhodosa **63**: 311-324.

Bateman, A. J. 1952. Self incompatibility system in Angiosperms I Theory, *Heredity* **6** : 285-310.

Bateman, A. J. 1954. The diversity of incompatibility system in flowering plants. *8th Congre. Int. Bot. Pairs. Rappcommsects* **9&10** : 138-145.

Batesmith, E. C. 1948. Paper chromatography of anthocyanins and related substances in petal extracts. Nature 161: 835.

Batesmith, E. C. and Metcalfe, C. R. 1957. Leuco-anthocyanins. The natural and systematic distribution of tannins in Dicotyledonus plants. *J. Linn. Soc. (Bot).* **55**:669-705.

Batesmith, E. C. and T. Swain. 1966. The asperuloids and the aucubins in:T. Swain (Ed). Comparative phytochemistry. *Academic Press. London. pp.* 159-174.

Batesmith, E. C. 1958. Plant phenolics as taxonomic guides. *Proc. Linn. Soc.* **169**:198-211.

Batesmith, E. C. 1962. Phenolic constituents of plants and their taxonomic significance. *J. Linn. Soc. Bot.* **58**:95-173.

Behnke, H. D. 1972. Sieve elements plastids in relation to angiosperm systematics. An attempt towards a classification of ultra structural analysis. *Bot. Rev.* , **38**:155-191.

Behnke, H. D. 1977. Transmission electron microscopy and systematics of flowering plants, In: K. Kubitzki (Ed) Plant systematics and evolution suppl I. *Springer Verlag. New York. Pp.* 155-178.

Bell, C. R. 1961. Phylogeny genetic variation and plant breeding. In : *P. J. Wanstall(ed). A Darwin centenary (B. S. B. I. Conference) Report No.* **6:** 121-129.

Bell, C. R. 1967. **Plant variation and classification**. *Walsworth publishing company. Belmonte.* California.

Bentham, G. &J. D. Hooker. (1862-1883). **Genera Plantarum**. Vols. I-III. Reeave & Co. , Williams & Norgate, London.

Bentham, G. 1873.

Bertarelli, E. 1902. il metodo biologico ele sue applicazione all diagnoi differenziable dille farine delle leguminose. *G. Acad. Med. Torino.* **8**:489-492.

Bessey, C. E. 1915. **The phylogenetic taxonomy of flowering plants.**

Bessey, C. E. 1894. **Evolution and classification**. *Contr. Bot. Dept. Univ. Nebraska.* NSF.

Blackslec, A. F. 1922. Variation in *Datura* due to changes in chromosome number. *Amer. Nat.* **56**:16-31.

Blackslec, A. F. 1930. Extrachromosomes, a source of variation in the Junson weed. *Smithosonian Report,* **1930**:431-450.

Blair, W. F. and B. L;Turner. 1972. The integrative approach to biological classification in: J. A. Behnke(Ed) Challenging

Biological Problems. *Oxford Univ. Press. New York. Pp.* 193-217.

Blair, W. F. , 1947. Estimated frequencies of the buff and grey genes.

Bobcock, E. B. & Stebbins, G. L. 1938. The American species of *Crepis*, their relationships, distribution as affected by polyploidy and Apomixis. *Carnegie Insti. Washington Publ. No. 504.*

Bock, W. J. 1959. Preadaptation and multiple evolutionary pathways. *Evolution*, **13**:194-211.

Bock, W. J. 1974. Philosophical foundations of classical evolutionary taxonomy. *Syst. Zool.* **22**:375-392.

Bohlman, F. , Hornowski, H. and Arndst, C. 1966. Occurrence of dihydro falcarinon in *Cotula coromopifolia*. Chem. Ber, 99: 2828

Boivin, B. 1967. Enumeration des plantes du Canada. III. Herbidees perpative Digitae: *Dimera, Liberae. Naturalista can:* **93**:583-646.

Bonnet, H. T and E. H. New Comb(1965). Polyribosomes and cisternal accumulations in root cells of raddish. *J. Cell. Biol,* **27**;423-432.

Borrill, M. 1961. The pattern of morphological variation in diploid and tetraploid *Dactylis. J. Linn. Soc. (Bot);* **56**:441-452.

Boulter, D. 1973. The use of comparative amino acid sequences data in evolutionary studies of higher plants. In:Progress in phytochemistry (L. Rienhold & Y. Liwschitz, eds.) vol 3. 199-222. Interscience . London.

Boulter, D. 1976. The evolution of plant proteins with special reference to higher plants cytochromes C. In: "Commentaries in Plant Science (H. Smith. ed.) pp79-91. Pregumam & Oxford.

Bouman, F. 1971. The application tegumentary studies to Taxonomic and phylogenetic problems. Ber. Dt. Bot. Ges. . **84**: 169-177.

Bowden, W. M. 1960. An experimental taxonomist examines the classification of the grasses. *Rev. Canad. Biol.* , **19**:279-392.

Boyden, A. 1964. Perspectives in systematic serology. In : C. A. Leon (Ed). Taxonomic Biochemistry and Serology.

Boyden, H. 1967. The place of precipitin testing among newer trends in taxonomy. *Bull. Nat. Inst. Sci, India.* **34**:108-117.

Bradshaw, A. D. 1962. The taxonomic problems of local geographical variation in plant species. *Systematic Assoc. Publ. No. 4. 7-16.* London.

Bradshaw, A. D. Population differentiation in *Agrostis tenuis*. I. Morphological differentiation. *New phyto,* **58**:208-227.

Brewbaker;J. L. 1967. The distribution and phylogenetic significance of binucleate and trinucleate pollen grains in the Angiosperms. *Amer. J. Bot.* **54**:1069-1083.

Brimacombe, R. 1981. Secondary structure and evolution of ribosomal RNA. Nature 94: 209-210.

Brunfels. O. Herbarium Vivae icons . Argentorati, 3 tomes in 1530.

Burrows, C. J. 1960. Studies in Pimelea I. The breeding system. *Trans Roy. Soc. New zeal.* **88** : 29-45.

Caesalpinus, A.

Camp, W. H, and C. L. Gilly, 1943. The structure and origin of species. *Brittonia* **4:** 323-385.

Carlquist, S. 1961. **Comparative Plant Anatomy**. Holt, Rinehart and Winston, New York.

Carlquist, S. 1975. Ecological strategies of xylem evolution. 259pp. University of California press, Berkeley, Los Angels, London.

Carlquist, S. 1981. Wood Anatomy of Pittospermaceae. *Allertonia.* **2**:355-392.

Chrispeels, M. J. and Baungartner, B. 1978. Serological evidence confirming the assignment of *Phaseolus aureus* and *P. mungo* to the genus *Vigna. phytochemistry.* **17**:125-126.

Clausen. J, and W. Hiesey 1958. **Experimental studies on the nature of species IV**. Genetic structure of ecological races.

Clausen. J, D. D. Keck, and W. Hiesey, 1940. Experimental studies on the nature of species I. Varied environments on Western North American plants. Carnegie. *Inst. Wash. Publ 544.*

Clausen. J, D. D. Keck, and W. Hiesey, 1947. Heredity of geographically and ecologically isolated races. *Ann. Nat.* **81**:114-133.

Clausen, J, D. D. Keck, and W. Hiesey, 1948. Experimental studies on the nature of species II. Environmental responses of climatic races of *Achillea.* Carnegie inst. Wash. Publ 581.

Cleland, R. E. 1924. The problem of species in *Oenothera. Amer. Nat.* 78:5-28.

Cleland, R. E. 1936. Some aspects of the cytogenetics of *Oenothera. Bot. Rev.* **2**:316-348.

Cleland, R. E. 1972. (*Oenothera*) **Cytogenetics and Evolution**. Academic press, London, New York. Cliffs. New Jersey.

Clevalein Lamarck, 1774- 1829. The philosophical utility of theory of Organic evolution.

Cleveland, L. R. 1949. The whole life cycle of chromosomes and their coiling systems. Trans. Am. Phil. Soc. 39: 1-1000.

Correns, C. 1928. Best. Mmung, Vererbung, and Verteillung des geschlechtes bie den hoheren pflanzen. Hand grab 2: 1-138.

Cowe. L. K. 1964. Evolution of out breeding in plants: I. The angiosperms *Heredity* **19**: 435-457.

Cronquist, A. 1965. The status of the general system of classification of flowering plants. *Ann. Missouri. Bot. Gard.* **52** : 281-303.

Cronquist, A. 1968. **Evolution and classification of flowering plants**. Nelson. New york.

Cronquist, A. 1969. On the relationship between taxonomy and evolution. *Taxon.* **18**:177-193.

Cronquist, A. 1977. On the taxonomic significance of secondary metabolites in Angiosperms. In:k. Kubitzki (ed) - plant systematics and evolution (Suppl. I) Springer-Verlag, New York. Pp. 179-189.

Cronquist, A. 1980. Chemistry in plant taxonomy: an assessment of where we stand. In : chemosystematics: principles and practice. (F. A. Bisby, J. G. Vaughan and C. A. Wrights, eds). Pp 1-27, Academic press.

Cronquist, A. 1981. **An integrated system of classification of flowering plants**. Columbia University Press. New York.

Crossby, J. L. 1949. Selection of an unfavourable gene complex: *Evol.* , **3**:212-230.

Curtis, J. H. 1959. The Vegetation of Wisconsin. *Univ. Wisconsis. Press.* Madison.

Dactylis. *J. Linn. Soc. (Bot.);* **56** : 441-452.

Dansear, B. H. 1950. A theory of Systematics Bible. *Biotheor. Serv. B.* **4**:117-189.

Darligton, C. D, and Shaw G. W 1958. Parallel polymorphism in heterochromatin of *Trillium* species. *Heredity,* **13**:89-121.

Darligton , C. D, and Wylie AP. 1955. **Chromosome Atlas of Flowering plants**. London.

Darlington, C. D. and La Cour. L. F. 1950. Hybridity selection in *Canapamula. Heridity.* **4**:217-249.

Darlington, C. D. 1958. **The evolution of genetic system**. 2nd ed. Basic Book, New York.

Darlington, C. D. 1963. **Chromosome botany and the origin of cultivated plants**. 3rd ed. Hafner, New York.

Darwin, C. 1859. **The origin of species by Natural selection**. London. 1876. The effects of cross and self fertilization in the vegetable kingdom.

Darwin. C. 1877. The different forms of flowers of plants by the same species, London.

Davidson, J. F. 1947. The polygonal graph for simultaneous portrayal of several variables in population analysis. *Madrono.* **9**:105-110.

Davis, P. H. and V. H. Heywood1963. **Principles of Angiosperm taxonomy**. Oliver Boyd. London.

Dawson, G. W. P. 1962. **An introduction to the cytogenetics of polyploids**. Oxford.

De Bary, A. 1884. **Comparative Anatomy of Vegetative organs of the Phanerogams and Ferns**. Clarendon Press. Oxford.

De Condle, A. P:1813. **Theorie elementaire**. Paris.

De Condle, A. P:1823-1873. Prodromus systematics naturalis regni vegetabilis. 17 vol and 4 index vols. Paris.

Dobzhansky, T. 1937. What is a species? *Scientia* **61** : 280-286. 1950. Mendelian populations and their evolution. *The Amer. Nat.* **84**:401-418.

Dobzhansky, Th. 1950. Mendelian populations and their evolution. *Amer. Nat.* **84**:401-418.

Dufou, O. 1951. Structure cytologique du pollen de quelques angiosperms. *Rev. Gen. Bot.* **68**:73-110.

Duval-Jouve, M. J. 1875. Histotaxic des feuilles de Gramineus. *Ameri. Sci. Nat. Bot.* **8**:227-246.

Eames, A. J. 1953. Floral Anatomy as an aid in generic limitation. *Chronica. Bot.* **14** : 126-132.

Eames, A. J. 1961. **Morphology of the Angiosperms**. McGraw Hill Book. Co, Inc. New York.

East, E. N. 1916. Studies in size inheritance in *Nicotiana. Genetics.* **1**:164-176.

Ehrendorfer, E. 1958. Zur phylogenic der gattung Galuino VI. Ein variability as zentrunals. Fossiler.

Ehrendorfer, E. , D. Schzier, H. Grger&C. Humphries. 1977. Chromosome banding and synthetic systematics in Anacyclus. (Asteraceae, Anthemidae). *Taxon.* **26**:387-394.

Eichler, A. W. 1875, 1878. Blulendragramme. 2 Vol. Wilhelm Engleman, Leipzig.

Endlicher, S. L. 1836-1840. **Genera Plantarum**. Vienna.

Engler, A. and K. Prantl, 1887-1915. Die naturlichen Pflanzenfamilien. 20 Vol. Wilhelm Engleman, Leipzig.

Epling, C. 1947 a. Natural hybridization of *Salvia apiana* and *S. mellifera. Evolution* **1**:69-78.

Epling, C. 1947 b. Actual and Potential gene flow in natural population. *Ameri. Nat.* **81** : 104-113.

Epling, C. and T. Dobzhansky; 1942. Micrographic races of *Linamthus_panyae.* **Genetics** **27**:319-332.

Erdtman, G. 1952. **Pollen morphology and plant taxonomy**. The Chronica Botanica. Co. Walthan.

Erdtman, G. 1960. Pollen walls and angiosperm phylogeny. *Botanical Notsser.* **113**:41-45.

Erdtman, G. 1963. Palynology. In R. D. Preston (Ed). *Advances in Botanical Research.* **1**: 149-208.

Ernst, A. 1953. Basic numbers Und polyploidy und ihre Bedewtung Fur dab. Heterostylie-problem. Archiev. *Julius Klaus Styfting,* **28** : 1-159.

Fair brothers, D. E. & Johnson, M. A. 1959. The precipitation reaction as an indicator of relationship in the subfamily Festucoideae of the family Poaceae (Graminae). *Proc. 9th Int. Bot. Congr. Montreal.* **2** : 110-111.

Farris, J. S. 1983. The empirical basis of phylogenetic systematics. In: "Advances in cladistics: proceedings of the second meeting of the willi Henning society" (N. Platnick and V. A. Fumkeds). Pp. 7-36. Columbia university press. New York.

Favali, M. A. and Gerola. 1968. Tubular and fibriollar components in phloem *Brassica chinensis. L.* Leaves. *Giom. Bot. Ital.* **102**:93-115.

Fisher, R. A. 1936. Has mendels work been rediscovered? *Ann. Sci.* **1** : 115-137.

Fowler, W. A. 1956. The origin of the elements. Sci. American. September 1.

Fox, D. J, Thurman, D. A. and Boulter, D. 1965. Studies on proteins of seeds of the Leguminosae. I. albumins. *Phytochem.* **3**:417-419.

Fryxell, P. A. , 1957. Mode of reproduction of higher plants. *Bot. Rev.* **23**:135-233.

Fryxell, P. A. , 1964. The interpretation of Distinct distribution. *Taxan* **16**:316-324.

Garber, E. D. 1958. The genus *Collinsia* III. The significance of chiasmata frequencies as a cytotaxonomic tool. *Madrono,* **14** : 172-176.

Gartner, K. F. 1849. **Versuche und Beobachting aber die Bastard-Zeugang** in Pflanzenreich - Stuttgart.

Gause, G. F. 1934. **The struggle for existence**. Baltimore.

Gilmour, J. S. L & Gregor, J. W. 1939. Demes: a suggested new terminology, *Nature.* **144**:333-334.

Gilmour, J. S. L. 1960. The deme terminology. *Report Scottish Plant Breeding Station.* **1960**: 99-105.

Gilmour, J. S. L. & Heslop-Hanison, J. 1954. The deme terminology and the units of microevolutionary change. *Genetica.* **27**: 147-161.

Goodwin, T. W. 1969. The carotenoids. In Swain. T. (ed). Comparative phytochemistry, pp. 121-137. Academic press. London, New York.

Goodwin, R. H. 1937a. Notes on the distribution and hybrid origin of *Solidago asperula. Rhodora.* **39**:22-28.

Goodwin, R. H. 1937b. The Cytogenetics of two species of *Solidago* and its beaming on their polymorphy in nature. *Amer, J. Bot.* **24**:425-432.

Gordon, A. D. 1981. **Classification**. Chapman and Hall, London.

Gottlieb, L. 1968. Hybridization between *Arctostaphylis viscida* and *A. cansescens* in oregon. *Brittonia.* **20**: 83-93.

Govinda Rajulu, E. 1969. **The Systematic anatomy of South Indian Cyperaceae**: *Cyperus L.* Subgenus *kylinga.*

Grant, K. A. and V. Grant, 1964. Mechanical isolation of *Salvia apiana* and *S. mellifera* (Labiatae). *Evolution.* **18**:196-212.

Grant, V. 1963. **The origin of Adaptations**. New york.

Grant, V. 1964. The biological composition of a taxonomic species in *Gilea. Adv. in. Geneti.* **12**:281-327.

Grant. V. 1952. Isolation and hybridization between *Aquilegia formosa* and *A. pubescens. Aliso.* **2**:341-360.

Gregor Mendel. 1865. **How are variations transmitted?**

Greilhuber, J. (1977). Why plant chromosomes do not show G. bands. *Theo. Appl. Genet.* **50**;121-124.

Greilhuber, J. (1982). Trends in the chromosome evolution of *Scilla.* (Liliaceae). *Stapfia.* **19**:11-51.

Gunderson, A. 1950. **Families of Dicotyledons**. Waltham, Mass, USA.

Gunthers, R. L. 1959. **The Greek herbal of disorders**. Hoftner publishing and co. New York.

Hair, J. B. and Beuzenberg, E. J. 1961. High polyploidy in New Zealand *Poa. Nature.* 189:160.

Hair, J. B. and E. J. Beuzenberg. 1958. Chromosomal evolution in the Podocarpaceae. *Nature.* **181**:1584-1586.

Hallier, H. 1912. Lorigine et le systeme phyletique des angiosperms. *Arch. Neerl. Sci.* **1**:146-234.

Harborne, J. B(1977). Flavonoids and evolution of the angiosperms- *Biochem. System. Ecol.* 5:7-22.

Harborne, J. B. and T. Swain(eds). 1969. **Perspectives in Phytochemistry**. Academic Press. New York.

Harborne, J. B. , D. Boulter and B. L. Turner 1971. **Chemo.axonomy of the Leguminosae**. Academic press. New York.

Harborne, J. B. 1966. *Comparative phytochemistry.* (T. Swain ed.) pp 271-295. Academic Press. London, New York.

Harborne, J. B. 1967. **Comparative biochemistry of flavonoids** 383pp. Academic Press. London, New York.

Harborne, J. B. 1970. **Phytochemical phylogeny**. Proceedings of the phytochemical society symposium-Bristol, April1969, London, Academic Press 1970.

Harlan Lewis and P. H. Raven, 1958a, *Clarkia franciscana* a new species from central California. *Brittonia.* **10** : 7-13.

Harlan Lewis and P. H. Raven, 1958b. Rapid evolution in *Clarkia. Evolution.* **12** : 319-336.

Harlan Lewis, 1961. Experimental sympatic population of *Clarkia. Am. Nat.* **45** : 155-168.

Harlan Lewis, 1962. Catastrophic selection as a factor in speciation. *Evolution.* **16**:257-271.

Harlan Lewis, D. 1942. The evolution of sex in flowering plants. *Biol. Res.* 497-501.

Harlan, J.-& D. E. Wet, 1963. The compilospecies concept. *Evolution.* **17**: 46-47.

Hedge, I. C. 1961a. Some remarks on the perennial species of *Ziziphora* with special reference to those in Turkey. *Notes Roy. Bot. Gard. Edinburgh,* **23**: 209-221.

Hedge, I. C. 1961b. Materials for a flora of Turkey : V. Notes Roy. *Bot. Gard. Edinburgh,* **23** : 547-558.

Hegde, I. C. 1961. Some remarks on the perennial species of *Ziziphora* with special reference to those in Turkey. *Notes. Roy. Boy. Gard. Edenbergh.* **23**:209-221.

Hegnaeuer, R. 1965. **Chemotaxonomic der Plangen Monocotyledoneae**. Basel&Stuttgart. Dicotyledons; Acanthaceae, Cyrillaceae. Basel&stuttgarrt.

Hegnaeur, R. 1966. *Comparative phytochemistry of Alkaloids.* T. Swain(Ed). Comparative phytochemistry. Academic Press. London. 211-230.

Hegnaeur, R. 1977. Cyanogenic compounds as systematic markers in Tracheophyta. *Pl. Sys. Evol. Suppl.* **1**:191-209.

Heilborn, O. 1929. Chromosome numbers and Taxonomy. *Proc. Int. Cong. Plant Sci.* **1**:307-310.

Heilborn, O. 1932. Aneuploidy and polyploidy in Carex. *Svnsk Bot Tidskr.* **26**:137-146.

Heilborne, O. 1939. Chromosome studies in cyperaceae. III, IV. *Heriditas.* **25**:224-240.

Heiser, C. 1947. Hybridization between the sunflower species

Helianthus annuus and *H. petiolaries. Evolution* **1**:249-262.

Heiser, C. B. , and W. C. Martin and DM. Smith. 1962. Species crosses in *Helianthus*: I. Diploid species. *Brittonia.* **14**:137-147.

Heiser, C. B. 1949. Study in the evolution of the sunflower species *Helianthus annuus* X *H. bolanderi. Univ. Cali. Publi. Bot.* **23**:157-208.

Heiser, C. B. 1951a. Hybridization in the annual sunflower *Helianthus annuus* **x** *H. debilis* var cucumerifolius. Evolution **1**: 249-262

Heiser, C. B. 1951b. The sunflower among the North American Indian Proc. *Am. Phil. Soc.* **95**:432-448.

Heiser, C. B. 1954. Variation and sub speciation in the common sunflower *Helianthus annuus. Am. Midl. Nat.* **51**:287-305.

Heiser, C. B. 1961. Natural Hybridization and introgression with particular reference to *Helianthus. Recent advances in Bot.* **1**:874-877.

Herr, T. M. Jr. 1971. A new cleaning technique by the study of ovule development in angiosperms. *Amer. J. Botany.* **58(8).** 755-790.

Herz, W. and M. Miyazaki, and Y. Kishida, 1961. Structures of parthenium and ambrosin. Tetrahedron letters. No. 2; 82-86.

Herz, W. 1968. In: **Recent Advances in Phytochemistry**. (T. J. Mabry, R. V. Alston and V. C. Runeckles. Eds) pp. 229-269. Appleton-Century Crofts, New York.

Heslop-Harrison, J. 1955. The Conflict of London categories. In:J. E. Lousley(ed.), Species studies in the British Flora : 160-172. *London. Bot. Soc Brit.*

Heslop-Makison, J. 1953. **New concepts in Flowering Plant Taxonomy**. Heinemann, London. educational Books(India). New Delhi.

Heywood, V. H. 1959. The taxonomic treatment of ecotypic variation. *Systematics Associ. Publ.* **3** : 87-112. London.

Heywood, V. H. 1971. **The Biology and Chemistry of the Umbelliferae**. Academic press London, New York.

Hilborne, O. 1936. Chromosome studies in Cyperaceae, III-IV. *Heriditas,* **25**:224-240.

Hillebrand, G. R and D. E. Fairbrothers. 1969. A serological investigation of intragenetic relationship *Viburnum*, caprifolaeceae. *Bull. Turr. Bot. el.* **96**:446-567.

Hillebrand, G. R and D. E. Fairbrothers. 1970. Phytoserological systematic survey of the Caprifoliaceae. *Brittonia.* **22**:125-133.

Hofmeister, L. 1956. Uber die Plasmagrenzschichten in Pollenkorn Protoplasma. **46**:367-379.

Holtman, R. E. 1955. Growth habits in monocotyledons, Variation theme. Phytomorphology. 5: 399-413.

Hooker, J. D. 1872. **The Flora of British India.**

Hooker, J. D. 1880. **Flora of British India.** 7 vols. London.

Hummock.

Hutchinson, J. B. 1959. **The families of Flowering plants** ed. 2. Oxford University press.

Hutchinson, J. B. 1969. **Evolution and phylogeny of flowering plants.** Dicotyledon: Facts & Theory. Academic press. London.

Hutchinson, R. 1973. The families of Flowering plants. Dicotyledons: Facts & Theory. Oxford university press. Oxford.

Huxley, J. S. 1938. *Clines*: an auxillary taxonomic principles, *Nature.* **142** : 219-220.

Huxley, J. S. 1942. Evolution: The modern synthesis. Allen and Unwin. London. 1955. Morphism and Evolution. *Heredity.* **9** : 1-52.

Huxley, J. S. 1939. Clines: an auxiliary method in taxonomy. *Bijdr. Dierk.* **27**: 491-520.

Hybrid Komplex, Der Oslmediterrane *Galuino graecem, L. G. canum.* Req. Former kreis. *Osterr Bot. Z,* **105**: 229-279.

Hyde, H. A & Williams. D. H

Hyde, H. A. and K. F. Adams. 1958. An atlas of airborne pollengrains. Mc Millan company. London.

Inamdar, J. A. 1968. Ontogeny of Stomata in some Oleaceae. *Proce. Indian Acad. Sci. Sect* :B67:157.

Irwin, H. S. and Rogers, D. J. 1967. Monographic studies in Cassia (Leguminosae, Caesalpinoidae) II. A taximetric study of section Apoucouita. *Mem. N. Y. Bot. Gard.* , **16** : 71-118.

Iverson, T. h. and P. R. Flood. 1970. The morphology, occurrence and distribution of diluted cisternae of the endoplasmic reticulum in tissues of plants of the Cruciferae. *Protoplasma.* **71**:467-477.

Jackson B. D. 1928. A Glossary of Botanical terms; Duck worth. London.

Jackson, R. C. 1962. Interspecific hybridization in *Haplopappus* and its beaming on chromosome evolution in the *Blepharadon* section. *Amer. Jour. Bot.* **49**:119-132.

Jackson, RC, 1957. The low chromosome number for plants. *Science.* **128**:1115-1116.

Jackson, RC, 1962. Interspecific hybridization in *Haplcpappus* and and its bearing on chromosome evolution in the *Blephardon* Section. *Amer. Jour. Bot.* **49**:119-132.

Jardine, N. and Sibson, R. 1971. **Mathematical taxonomy**. John Wiley, London.

John Ray, (1682). Methods Plantarum Nova.

John Ray, (1686). Historia Plantarum, Dicotyledons: Acanthaceae, Cyrillaceae. Basel & Stuttgart.

John Ray, 1703. Methods Plantarum. Nova.

Jones, 1961. The status and development of abrupt ecospecies. *Recent Advances in Botany.* 862-866. Toronto Univ. Press.

Jones, C. E. and R. J. Little (eds). 1983. **Hand book of Experimental Biology Scientific and Academic editions**. *Van Nostard Reinhold Company. Inc.*, New York.

Jones, D. F., 1932. Interaction of specific genes determining sex in dioecious maize. *Proc. 6th Intern. Congr. Genetics N. Y. Ithaca.* **2**:104-107.

Jones, W. H. 1970. Scanning electron microscopy of pollen as an aid to the systematics of *Vernonia*(Compositae). *Bulletin. Tommey. Botanical Club.* **97**:325-335.

Jordan, A. 1864. Diagnoses diespeces nouvelles ou meconnues. Paw.

Jorgensen, C. A. Sorensen, T. & Westgaurd, M. 1958. The flowering plants of Greenland, a taxonomical and cytological survey. *Biol. SKR. Dan. Vid. Selsk.*, **9**:1-172.

Jorgensen, C. A. Sorensen, T & Westgaard, M. 1958. The flowering plants of Greenland, a taxonomical and cytological survey. *Biol. SKR. Dan. Vid. Selsk.*, **9**:1-172.

Kajale, L. B. 1942. A contribution to the embryology of the genus

Portuluca. J. Indian. Bot. Soc. **21**:1-19.

Kale and Pai.

Kjaer, A. 1960. Naturally derived isothiocynalies (mustard oils) and their present glucosides. *Progress in the chemistry of organic natural products.* **18**:126-176.

Kliephius. E. 1967. Cytotaxonomic notes on some *Galium* species. *Act. Bot. Nurl.* **15**:535-538.

Kooiman. P, 1972. The occurrence of iridoid compounds - their occurrence and systematic importance in angiosperms. *Bot. Notiser.* **128**:148-180.

Kruckberg, A. R. 1951. Intraspecific variability in the response of certain native plant species to Serpentine Soil. *Amer. Jour. Bot.* **38**: 408-419.

Kruckberg, A. R. 1954. The ecology of Serpentine Soils III. Plant species in relation to serpentine species. *Ecology.* **35**:267-274.

Kruckberg, A. R. 1957. Variation in fertility of hybrids between Isolated Populations of the Serpentine species, *Strepanthus glandulosus. Hook. Evolution.* **11**:185-211.

Kruckberg, A. R. 1958. The taxonomiy of the species complex *Strepanthus glandulosus. Hook. Madrono.* **14**:217-248.

Kruckberg, A. R. 1969. The implication of Ecology for plant systematics. Taxon 18 : 185-211.

Kyhos, D. W. 1958. M. , 1958. the independent aneuploid origin of two species of *Chaenactis* (Compositae) from a common ancestor. *Evolution.* **19**: 26-43.

Ladizinsky, G. and Hymowrtz, T. 1979. Seed protein electrophoresis in taxonomic and evolutionary studies. Theor. Appl. Genet. 54 : 145-151.

Lawrence G. H. M, 1951. **Taxonomy of vascular plants.** The McMillan Company, New York

Lawrence G. H. M. 1955. **An introduction to plant taxonomy**. The Mac Millon company New York.

Lawrence, G. H. M. 1965. **Herbals their history and significance**. In: History of Botany, 2 papers presented at a symposium held at the William Andrew Clerk Memorial Library Dec 7th 1963, University of California Los Angels.

Lerner, M. 1958. **The Genetic Basis of Selection**. New York.

Lesquereux, L. 1891. **The Flora of Dakota group**. *U. S. Geol. Surv. Mon.* **17**:1-400.

Lester, R. N. 1965. Immunological studies on the tuber bearing solanums. Techniques and south American species. Ann. Bot. 29: 609-624.

Lewis H, P. H. Raven, (1958) Rapid evolution in *Clarkia. Evolution*, **12**: 319-336.

Lewis, D & Crowe, L. K. 1956. the genetics and evolution of Gynodioecy. *Evol.* , **10**: 115-125.

Lewis, D. & Crowe, L. K. 1958, Unilateral interspecific incompatibility in flowering plants. *Heredity*, **12**: 232-256.

Lewis, D. 1942. Comparative incompatability in angiosperms and fungi. *Adv. Genetics*. **6**:235-285.

Lewis, D. 1954. Comparative incompatibility in angiosperms and fungus. Adv. Genetics. 6 : 235-285.

Lewis, H. & P. H. Raven. 1858a. *Clarkia franciscana* a new species from Central California.

Lewis, H. & P. H. Raven. 1858b. Rapid evolution in *Clarkia. Evolution* **12**:319-336.

Lewis, H. 1962. Catastrophic selection as a factor in speciation. *Evolution*. **16**:57-71.

Lindsey, D. W. and R. K. Vickey, 1967. Comparative Evolution in *Mimulus guttatus* of the *Bonneville basin. Evolution*. **21**:430-456.

Lindsey, A. A. 1938. Anatomical evidence for the Menyanthaceae. *Amer. Jour. Bot.* , **25**:480-485.

Lindsey, A. A. et al, 1969. Natural Area in Indian and their Preservation. Amer. Mid. Naturalist press. Nature Dane Indian. 594pp.

Linnaeus, C. 1737. Critica Botanica; C. Wishoff. Amsterdam.

Linnaeus, C. 1737b. Flora rappenica; C. Wishoff. Amsterdam.

Linnaeus, C. 1751. Philosophia Botanica. Stockholm.

Linnaeus, C. 1753. Species Plantarum, Vols. 1&2, L. Salvi. Stockholm. 5th Edition.

Linnaeus, C. 1766-1790. Resquisitio de sexu plantarum. St. Petersberg. 1760. *Ammoenitates Academica Erlangen*. 1790.

Linnaeus, C. 1767. **Genera Plantarum**.

Linsbaver K. , Die Epiderms, in K. Linsbaver(Ed.)1930. Handsbuchder, Pflanzenantomic; Gebruder Borntraeger, Berlin, pp27.

Love, A. &Love, D. 1961. Chromosome number of Central and Northwest European Plant species. Opera Botanica Lund, S.

Lysenk, O. & Sneath, P. H. A. 1959. The use of models in bacterial classification. *J. Gen. Microbiology.* ;**20**:284-289.

Mabry, T. J. , R. E. Alston and V. C. Runackles (eds). 1968. **Recent Advances in Phytochemistry vol. I. Appletion Century-Crofts**. New York.

Mabry, T. J. 1966. **The Betacyanins and Betaxanthins in:T. Swain(Ed). Comparative Phytochemistry**, Academic Press, London, pp 231-244.

Mabry, T. J. 1976. Pigment dichotomy and DNA, RNA hybridization data for Centrospermous families. *Pl. Syst. Evol.* **126**:79-94.

Macior;L. W. 1971. Co-evolution of plants and animals - systematic insights from plant insect interactions. *Taxon* **20**:17-28.

Macior, L. W. 1977. The Pollination ecology of *Pedicularis* (Scropilanaceae) in the Science Nevader of California. *Bull. Torney. Bot. Club* **104**:148-154.

Magnus, W. 1908. Westere Ergebriss der serum-diagnostik fur du theoretische umd angewandte Botanik. *Ber. dt. bot. Ges* **26**:532-539.

Maheswari Devi, H and Pullaiah, T. 1975. Life history of *Basella rubra. L.* and taxonomic status of the family Basellaceae. *Jour. Ind. Bot. Soci.* **54:**154-166.

Maheswari Devi, H. 1962. Embryological studies in Gentianaceae (Gentianoideae) and Menyanthordeae, *Proc. Indian Acad. Sci.* B. **56**:195-216.

Maheswari, P. and R. N. Kapil. 1963. The morphological and embryological studies in the Lemnaceae I. The floral structure and Gametophytes of *Lemna paucicostata. Amer. Jour. Bot.* **50**:677-686.

Maheswari, P. (1950). *An introduction to Embryology of Angiosperms.* Mc Graw-Hill, New York.

Maheswari, P. 1964. *Embryology in relation to taxonomy.* Bull. Natl.

Insti. Sci. Ind, 34.

Maheswari, Sc. &Baldev, B. 1958. A contribution to the morphology and embryology of *Commelina_forkalaei*. Vohl. Phytomorphology, 277-298.

Manske, R. H. F, 1944. The Alkaloids. Vols I-VII.

Massart, J. 1902. Laccomadation individuelle chez le *Polygonium amphibiumn. Jord Bot Etat. Bruxelles* **1**:73-95.

Mather, K. 1946. **Statistical analysis in Biology,** 2nd edn, Chapman Hall, London.

Mather, K. 1949. **Biometrical Genetics**. New York. Dover Publication. 55p.

Mather, K. 1966. Breeding systems and response selection in reproductive biology and taxonomy of vascular plants. Ed. J. G. Hawkes. Pergamon press, Oxford.

Mathiessen, A. 1962. "A contribution to the embryology of *Paoenia*". *Act. Hort. Berg.* **20**:57-61.

Mayr, E. 1942. **Systematics and Origin of species**. New York.

Mayr, E. 1957. **Species concepts and definitions**. In the species problem. E. Mary(ed). Washington.

Mayr, E. 1959. **Trends in Avian systematics**. *Cold. spring. Herb. Symp. Quant. Biology.* **101**:293-302.

Mayr, E. 1966. **Numerical Phenetics and Taxonomic Theory**. *Syst. Zool.* **15:**73-97.

Mayr, E. 1970. **Populations, Species and Evolution**. Belknap press of Harvard University press. Cambridge. Mass.

McClintok, B. (1929). A cytological and genetical study of triploid maize *Genetica,* **14**:180-222.

McKelny, S. D. & Sax, K. 1933. Taxonomic and cytological relationship of *Yucca* and *Agave. J. Arn. Arb.* , **14**:76-81.

McMillan, 1959. The role of ecotypic variation in the distribution of the central grassland of North America. *Ecol. Monogr.* **29**:285-303.

McMIllan, C. 1956. Ecotypes and Community function. *Amer. Nat.* , **94**:245-255.

McMillan, C. 1959. The role of ecotypic variation in the distribution of the Central grassland of North America. *Ecol. Monogr.* **29**:285-308.

McMillan, C. 1964. Ecotypic differentiation within four North American prairie grasses:II. Morphological variation within transplanted Community functions. *Amer. Jour. Bot.* **5**:1119-1128.

McMillan, C. 1965. Ecotypic differentiation within four North American prairie grasses:II. Behavioral variation within transplanted community functions. *Amer. Jour. Bot* **52**:55-65.

McMillan, B. H. 1968. A method of mounting herbarium in liquid plasties; New Zealand *J. Bot.* **6**:514-517.

McNeill, J. 1975. A generic revision of Portulaceae tribe Montriceae using techniques of numerical taxonomy. Canad. J. Bot. 53: 789-809.

Meeuse, A. D. J. Interpretative gynoecial morphology of the Lactoriadaceae and Winteraceae-a reassessment. *Act. Botanica Neerlandiea.* **20**:221-238.

Melchior, H. (Ed.) (1964) AHG. **Engler's syllabus der Pflanzenfamilien** 12th Edition **Vol.** 2.

Metcalfe, C. R&Chalk 1950. **Anatomy of Dicotyledons**. Oxford. 243-301.

Metcalfe, C. R. 1954. Recent advances in taxonomy as seen by anatomist. *Congr. Inst. Bot. Paris. Rapp&Comm. sects.* **2, 4;5&6**:37-43.

Metcalfe, C. R. 1961. The anatomical approach to systematics. General introduction with special reference to recent work on Monocotyledons. In : *Recent advance in Botany*: 146-151 Toronto University, Press.

Mohinder Pal, 1962. The diploid and Tetraploid race of *Pupalia lappaceae, Juss Curr. Sci* **2**:346-347.

Mooney, H. F. 1950. **A supplement to the Botany of Bihar and Orissa**. Govt. Press. Ranchi.

Muller, C. H. ;1952. Ecological control of hybridization in Quercus: a factor in the mechanism of evolution. *Evolution* **6**:147-161.

Muller, C. H;1951. The Oaks of texas. Contrib. *Texas Research Foundations.* **7(3):**21-323.

Muller, J. 1970. Palynological evidence on early differentiation of angiosperms. *Biol. Rev.* **45**:417-450.

Murty, Y. S. and V. Puri. , 1980. Floral anatomy in relation to taxonomy. *Glimp. Pl. Res.* **5**: 222-243.

Naik, V. N. 1976a. Chromosome behavior and evolutionary trends in class *Chlorophytum* (Liliaceae). *Bot. J. Linn. Soc.* **72**:45-50.

Naik, V. N. 1976b. Cytotaxonomic studies in two species of *urginea* steinth from India. *J. Indian. Bot. Soc.* **55**:62-64.

Nair, P. K. K. 1974. The phylogeny of angiosprms – A palynological analysis. *Evol. Biolo.* **7**:127-136.

Newcomb, W. 1973. The development of the embryo sac of sunflower *Helianthus annuus* before fertilization. *Can. J. Bot.* **51:**863-878.

Ornduff, R. 1969. Reproductive biology in relation to systematics. Taxan 18:121-133.

Ornduff, R. 1976. Concepts of Species. (Book Review). *Syst. Bot.* **1**:347.

Paliwala, G. S, 1966. Structure and Ontogeny of Stomata in some Caryophyllaceae. *Phyto morphology,* **16**:533-539.

Paliwala G. S. (1968). Ontogeny of stomata in Cruciferae. *Cand. J. Bot.* , **45**:495-500.

Pandey, K. K. , 1960. Evolution of gametophytic and sporophytic systems of self incompatibility in angiosperms. *Evolution* **14**:98-115.

Pant, D. D and B. Mishra. 1965. Ontogeny of stomata in Rubiaceae. *Phytomorphology* **15**:300-310.

Pant, D. D and P. F. Kidwi 1964. On the diversity in the development and organization of stomata in *Phyla nodiflora. Michx. Curri. Sci.* , **33**:653-654.

Pant, D. D. 1965. On the Ontogeny of stomata and other homologous structures. *Pl. Sc. Ser.* **1**:1-24.

Paris, R. &Mysore, H. 1957. Examen en electrophorese sun papier des alcalordes totanx. De diverses especies de perkenches. C. R. *Acad. Sci.* , **245**:1265-1268.

Paris, C. D. W. J. Hanney and G. B. Wilson 1960. A survey of the interactions of genes for flower colors. Mich. State. Uni. Agri. Exp. Stat. Tech. Bull. 291.

Paris, R. R and Letouzey, 1960. Distribution of alkaloids in yohimbe. Jour. Agri. Trop. Bot. Appl. 7 : 256-258.

Paris, R. R. and Paris, M. 1964. C. r. hebd. seanc. *Acad. Sci.* Paris.

258, 361.

Payers J. P. D. W. 1967. A monograph of the genus *Baringtonia* (Lecythidaceae). *Blumea* **15**:157-263.

Payne, W. W. (1976). Biochemistry and species problems in *Ambrosia*. Asteraceae-Ambrosieae. Pl. Syst. Evol. 125: 169-178.

Payne, W. W. (1979). Stomatal patterns in embryophytes; then evolution, ontogeny and interpretation, *Taxon*. **28:**117-132.

Payne, W. W. 1970. Heliocyte and allelocytic stomata. Unrecognized patterns in the Dicotyledons, *Amer. J. Bot.* **57**:140-149.

Paynee, W. W, PH. Raven and D. W. Kyos(1964) chromosome numbers in compositae-IV. Ambrosieae; *Amer. J. Bot.* **51(4)**:419-424.

Percival, M. S. (1965). "**Floral Biology**". Pergamon Press. Oxford.

Perrot, E, 1897.

Pichon, N. 1946. Sur les Alismataces et les Butomacees. *Not. Syst. Paris* **12**:170-183.

Pickering, J. L&D. E. Fairbrothers. 1970. A serological comparison of Umbelliferae sub families. *Amer. J. Bot.* 57:988-992.

Plouvier, V. 1954. Surla presence du pinitol chez less caryophyllace set qudques plantes de familles. Voisines. Compt. Rend. Acad. Sciparis. **239**: 1678-1680.

Poddubnaya-Arnoldi(1976).

Prazmo, W. 1960. Genetic studies on genus Aquilegia. II Acta Folsk To Varz. Bot 29: 57-77. Bot 30: 421-442.

Puddubnaya-Arnoldi, V. A. 1959. Study of fertilization and embryogenesis in certain angiosperms. Using twing material. *Amer. Nat.* **93**:161-169.

Pullaiah, T. 1981. Studies in the Embryology of Heliantheae (Compositae). Plant. Sys. Evol. 137, **3**: 203-214

Pullaiah, T. , Laxminarayana K. & Hanumantha Rao B. 2001. Text Book of Embryology of Angyosperms, Regency Publications.

Puri, V 1947. Studies in Floral Anatomy IV Vascular anatomy of the flower of certain species of the Passifloraceae. *Am. J. Bot.* **34**:562-573.

Puri, V 1958. Floral anatomy and taxonomy. *Indian. bot. Soci. Mem.* **1**:15-18.

Puri, V. 1952. Floral anatomy in relation to taxonomy. *Agra. Uni. J. Res(Sci).* **1**:15-35.

Puri, V. 1962. Floral Anatomy in relation to taxonomy. *Bull. Bot. Survey, India.* **4**:161-165.

Radford, A. E. , Dickson, W. C. , Massey, J. R. and Bell, CR 1974. "**Vascular plant systematics**". Harper and Row, New York, London.

Radford, Albert. E. W. C. 1976 Dickinson and J. R. Massey and CR. Bell, 1976. "**Vascular plant systematics.** " New York and London.

Radhakrishnaiah, M. 1996. **Essentials of plant taxonomy**. Hyderabad.

Ramayya, N. &T. Rajagopal, 1968. Foliar epidermis as a taxonomic aid in the flora of Hyderabad. Part I. Portulucaceae and Aizoaceae. *J. Osmania Univ,* **4**:147-160.

Ramayya, N. &T. Rajagopal, 1971. Foliar dermotypes of Indian Aizoaceae and their use in identification. *J. Indian. Bot. Soc.* **50**:355-362.

Ramayya, N. 1969. The development of trichomes in the Compositae in : K. A. Chowdary (Ed). Recent advances in the anatomy of tropical seed plants. *Hindustan Publ. Congr. New Delhi.* pp85-113.

Raven, P. H. , and D. W. Kyhos. 1965. New evidence concerning the original basic chromosome number of angiosperms. *Botanical Review.* **18**:603-651.

Raven, P. H. 1975. The bases of angiosperm phylogeny. Cytology. *Ann. Miss. Bot. Gard.* **62**:724-764.

Rendel, A. B. 1930. **The classification of flowering plants** ed-2- Cambridge.

Rendel;A. B. (1904-1925). **The classification of flowering plants**. Vol. 1. Gymnosperms and Monocotyledons. 403pp. Cambridge 1904. Vol. 2. Dicotyledons 636pp. Cambridge, 1925. (The second edition of vol. 1. (1930) entailed only typographical changes and a brief appendix.)

Rhoades, M. M. &Mc clintock, B. 1935. The cytogenetics of *Maize. Bot. Rev.* **1**:292-325.

Rick, C. M. and Barton, D. W. 1954. Harper and Row. Cytological and genetcal identification of the primary trisomic of the tomato. *Genetics,* **39**:640-666.

Riley, H. P. & Bryant, T. R. 1961. The separation of nine species of the Iridaceae by paper chromatography. *Amer. J. Bot.* **48**:133-137.

Riley, R. 1956. The influence of the breeding system on the genecology of *Thlaspe alpestre. L. New Phytol.* **55**:319-330.

Rogers, A. D. III. 1942. John Torney. Princeton university press. Princeton. Reprinted 1965 by Hafner. New York. (The standard Bibliography of Torney.)

Rogers. D. G. & Tanimoto, T. T. 1960. A computer programme for classifying plants. *Science.* **132**:1115-1118.

Rollins, R. C. 1944. Evidence of natural hybridity between guayule (*Parthenium argentatum*) and mariola (*Parthenium minimum*). *Amer. J. Bot.* **31**:93-99.

Rollins, R. C. 1945. Interspecific hybridization in *Parthenium. I.* cross between guayule (*Parthenium argentatum*) and mariola (*Parthenium minimum*) *Amer. J. Bot.* **32**:395-404.

Rollins, R. C. 1946. Interspecific hybridization in *Parthenium* II. Crosses involving *P. argentitatum, P. stramonium, P. tomentossum, P. hysterophores. Amer. J. Bot.* , **33**:21-30.

Rollins, R. C. 1953. Cytological approaches to the study of genera *Chronica Botanica.* **14**:133-139.

Rudorf-Lauritzen, M. 1958. The trisomics of *Antirhinium majus* . L. *Proc. 10th. Int. Congr. Genet.* **2** (Abstracts):243-244.

Sahasrabudhe, S. and Stace, C. A. (1974). Developmental and structural variation in the trichomes and stomata of some Gesneriaceae. *New. Botnist,* **I**:46-62.

Satyavathi, D. & U. R. Murti 1974. The aneuploid complex of *Iseilema Anders* III. Artificial polyploids. *Nucleus.* **17**:76-79.

Schweizer, D. and F. Ehrendorfer, 1976. Geimsa banded Karyotypes systematics and evolution in anacyclus (*Asteraceae, Anthemideae*). *Pl. Syst. Evol.* **126**:107-148.

Shulz, O. E. 1936. Cruciferae in A. Engler, Dienaturlichen Pflanzenfamilien, 2nd Edn. **17b**:227-658.

Simpson, G. G. 1961. "**Principles of animal taxonomy**". New York.

Simpson, G. G. 1963. "**The major features of evolution**". Columbian University Press. New York.

Singh, R. S. , Jain, S. K. and Quadrat C. O. 1973. Protein electrophoresis as an aid to oat variety identification. Euphytica. 22: 98-165.

Singh, V, D. K. Jain and Meena Sharma 1978. Leaf architecture in Barberidaceae and its Bearings on the circumscription of the family. *J. Indian. Bot. Soc.* **57**:272-281.

Sinnot, E. W. 1914. Investigations on the phylogeny of the angiosperms. The anatomy of the Node as an aid in the classification of angiosperms. *Amer. J. Bot.* **1**:303-322.

Smith, H. H. 1943. Effects of genome balance, polyploidy and single extra chromosomes on size in *Nicotiana. Genetics,* **28**:227-236.

Sneath, P. H. A. , and R. R. Sokal. **"The principles and practice of numerical classification"**. W. H. Preemem Company. San Francisco.

Sneecor, G. W. , and W. G. Cochram. 1967. **Statistical methods**. 6th ed. Iowa state university press. Ames. Iowa.

Sokal, R. R. and P. H. A. Sneath. 1963. **Priciples of numerical taxonomy**. W. P. Preeman Company, San Francisco.

Solbrig, O. T. 1964. Intraspecific variation in the *Gutirrezia surrothrae* complex (Compositae - Asteraceae). *Contr. Grey. Herb.* **193**:67-115.

Solbrig, O. T. 1968. Fertility, Sterility and the species problem. In modern methods in plant taxonomy, V. H. Heywood(Ed) London.

Sorenson, T. 1958. Sexual chromosome aberrations in triploid apomictic. Taraxea. Bot. Tidsskar. 54: 1-22.

Soueges, R. 1907. Development et structure du tegmnert seminal chez les Solanace. *Ame. Sci. Nat. Bot. XII.* **6**:1-124.

Soueges, R. 1953. **Embryogenic des Peganacedes. Developnende. 1' embryon chez le-Peganum harmale** L. C. R. *Acad. Sci.*

Sperlich, A. (1939). Das tropicsche parenchym, B Exkretionsgewebe in : K. Linsbauer(Ed). **Handbuch der pflanzenanatomic Gebruder Borntraeger**, Berlin, P. 38.

Stace, C. A. 1965. Cuticular studies as an aid in plant taxonomy. *Br. Mus. (Nat. Hist.) Botany bulletin.* **4**:1-78.

Stace, C. A. 1966. By use of epidermal characters in phylogenetic considerations. *New Phytol.* **65**:304-318.

Stace, C. A. 1973. The significance of the leaf epidermis in the taxonomy of the *Combretaceae IV*. The genus *Combretum* in Asia. *Bot. J. Linn. Soc.* **66**:97-115.

Stace, C. A. 1974. The significance of Leaf epidermis in the taxonomy of combertaceae IV. The genes combretum in Asia. *Bot. J. Linn.* **66**:97-115.

Stebbins, G. L. , Genkins, T. A and Walters, M. S. 1953. Chromosomes and phylogeny in the Compositae tribe *Cichoreae. Univ. Calif. Publi. Bot.* **26**:401-430.

Stebbins, G. L. 1950. **Variation and evolution in plants**. Columbia Univ. Press. New York.

Stebbins, G. L. 1959. **Genes, Chromosomes and evolution**. In: Vistas in botany. W. Turril(Ed) London.

Stebbins, G. L...Aridity as an evolutionary stimulus. *Amer. Nat.* , **8**:33-44.

Stevens, R. A. and Martin, E. S. 1978. A new Ontogenetic classification of stomatal types. *Bot. J. Linn. Soc.* **77**:53-64.

Swarnson, L. T. 1958. **Cytology and Cytogenetics**. London.

Swartz, D. 1971. **Collegiate dictionary of Botany**. Ronald Press.

Takhtajan, A. L. 1969. **Flowering plants: Origin and Dispersal**. Oliver Boyd. Edinburg (English translation by C. Jeffrey.).

Takhtajan, A. L. 1980. Outlines of the classification of flowering plants (Magnoliophyta) *Bot. Rev.* **46(3)**:225-359.

Theiman, E. 1956. The cytotaxonomy of the tribe polygonatae. *Amer. J. Bot.* **49**:134-142.

Theobald, W. L. 1966. The Lomatum dasyi carpum *Mohavense-foeniculaceum* complex (Umeblliferae). *Briltonia,* **18**: 11-18.

Theobald, W. L. Krhulik, J. L. and Rollins, R. C. 1979, 1980. Trichome description and classification. In: "**Anatomy of Dicotyledons**". (C. R. Metcalfe and L. Chalk. eds. 2nd edn.) Vol 1. pp. 40-53. Clarendon press. Oxford.

Theophrastis, 1916. **Enquiry into plants,** translated by Sir Arthur Hort. Lieb. Classics Library, London. Harvard University Press. Cambridge Massachusetts, 1948-1949.

Theophrastis. 1529. **Historia et cansis plantarum**... T. Gaza interprete. Parisis.

Thorne, R. F. 1963. Some problems and guiding principles of

Angiosperm phylogeny. *Amer. Nat.* **97 (896):**287-305.

Thorne, R. F. 1968. Synopsis of a relatively phylogenetic classification of flowering plants, *Aliso*, **6**:57-66.

Thorne, R. F. 1976. A phylogenetic classification of the angiospermae in : M. K. Hecht, W. C. Steere and B. Wallace (Ed). Evolutionary biology Vol. 9, Plenum Press, New York and London. pp35-106.

Thorne, R. F. 1981. Phytochemistry and Angiosperm phylogeny. D. A. Young and D. S. Seigler, eds. pp. 233-295. Praeger, New York.

Tippo, O. 1938. Comparitive anatomy and morphology of the Moraceae and their presumed allies. *Bot. Gaz.* **100**:1-99.

Togby, H. A. 1943. A cytological study of *Crepis fuliginosa* , *C. neglecta* and their F1 hybrids and its Bearing on the mechanism of Phylogenetic reduction in chromosome number. J. Genetics. **45**:67-111.

Tournefort, J. P. , De, 1624. **Elements de Botanique**. . 3 vol. Paris.

Turesson, G. 1926. The plant species in relation to habitat and climate. *Heriditas* **6**:147-236.

Turner and J. B. Harborne, 1967. Distribution of canavanine in the Plant Kingdom. *Phytochemistry*, **6**:863-866.

Valentine, D. H&Love, A 1958. Taxonomic and biosystematic categories *Brittonia* **10**:153-166.

Van Cathem, 1970. A classification of stomatal types. *J. Linn. Soc. Bot.* **63**:235-246.

Van Steenis, CGGJ. 1957. Specific and infraspecific delimitation. Flora Malesiana Set 1, 5(3) cl XVII- CCXXIX. Djakarta.

Van Steenis, C. G. G. J. 1970. A review of A. Takhtajan **Flowering plants: Origin and dispersal**. *Blumica* **18**:222-224.

Van Valen, L. 1976. Ecological species, multi species and Oaks. *Taxon* **25**:233-239.

Van Valen, L. 1978. Contribution to the knowledge of cynogenesis in angiosperms 3. Communication. Cyanogenesis in Liliaceae Proc. Komnk. Nederl. Akad. Wetensch. Amsterderm. *Ser. C.* **81**:130-140.

Van Valen, L. 1978. Why not be a cladist. *Evol. Theory.* **3**:285-299.

Vancampo, M. 1976. Patterns of pollen morphological variation within taxa. In: The evolutionary significance of the Exine(I. K. Fergnson&Muller. eds) *pp*-125-138. Academic press, London, New York.

Vij, S. P & S. K. Kashyap. 1978. Palynological studies in some Orchidaceae. Abs. II. Ind. Palyn. Conf. Banglore. 22-23.

Wagner, W. H. Jr. 1970. Biosystematics and evolutionary noise. *Taxon* **19**:146-151.

Wagner, W. H. Jr. 1980. Origin and philosophy of the ground plan divergence method of cladistics. *Syst. Bot.* **5**:173-192.

Wallace, A. R. 1889. Darwinism: **An exposition of the theory of Natural Selection**. Mc millan, London.

Watson, J, D. and J. Clausen. 1961. Phenotypic responses to contrasting environments in the genus *Poa*. Scottish plant breeding station report.

Wegener, A (trans. skebl) J. G. A. **The origin of Continents and Ocean**. London, 1924.

Weiseman, H. 1872. "**Uber den Einfluss der Isoiereng auf die Artbildung**". Wilehm-Engelmann. Lei P319.

Weiseman, H. 1909. **The selection theory**. In: Darwin and modern science ed, A. C. Seward. Cambridge University press. Cambridge.

Whitehouse, H. L. K. 1950. Multiple allenomorphic incompatibility of pollen and style in the evolution of the angiosperms. *Ann. Bot. (N. S)* **14**:199-216.

Wiley, E. O. (1981). Phylogenetics, "**The theory and practice of phylogenetic systematics**" John Wiley, Brishbane.

Wilkins, D. A. 1960. **Recognising adaptive variants**. *Proc. Linn. Soc. Lond.* , **171**:122-126.

Williams, A. H. 1966. Dihydro chalcones. In : T. Swain (ed). Comparative phytochemistry, pp. 297-307. Academic press. London and New York.

Williams, G. C. 1966. **Adaptation and Natural selection**. Princeton University Press. Princeton, New Jersey.

Williams, W. T and Lambert, J. M. 1961. Multi variate methods in taxonomy. *Taxon*. **10**:205-211.

Williams, W. T. 1967. The Computer Botanist. Australian Journal of Science. **29**:266-271.

Williams, W. T. 1969. The problem of attribute weighting in numerical classification. *Taxon* **18**:369-374.

INDEX

A

B

L

M

N

Q

R

S

T

U

V

W

Z